Preservation of random megascale events on Mars and Earth: Influence on geologic history

Edited by

Mary G. Chapman
and
Laszlo P. Keszthelyi
Astrogeology Team
U.S. Geological Survey
2255 N. Gemini Drive
Flagstaff, Arizona 86001
USA

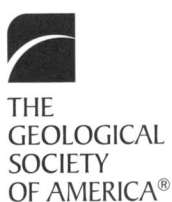

THE
GEOLOGICAL
SOCIETY
OF AMERICA®

Special Paper 453

3300 Penrose Place, P.O. Box 9140 • Boulder, Colorado 80301-9140 USA

2009

Copyright © 2009, The Geological Society of America (GSA). All rights reserved. GSA grants permission to individual scientists to make unlimited photocopies of one or more items from this volume for noncommercial purposes advancing science or education, including classroom use. For permission to make photocopies of any item in this volume for other noncommercial, nonprofit purposes, contact The Geological Society of America. Written permission is required from GSA for all other forms of capture or reproduction of any item in the volume including, but not limited to, all types of electronic or digital scanning or other digital or manual transformation of articles or any portion thereof, such as abstracts, into computer-readable and/or transmittable form for personal or corporate use, either noncommercial or commercial, for-profit or otherwise. Send permission requests to GSA Copyright Permissions, 3300 Penrose Place, P.O. Box 9140, Boulder, Colorado 80301-9140, USA. GSA provides this and other forums for the presentation of diverse opinions and positions by scientists worldwide, regardless of their race, citizenship, gender, religion, or political viewpoint. Opinions presented in this publication do not reflect official positions of the Society.

Copyright is not claimed on any material prepared wholly by government employees within the scope of their employment.

Published by The Geological Society of America, Inc.
3300 Penrose Place, P.O. Box 9140, Boulder, Colorado 80301-9140, USA
www.geosociety.org

Printed in U.S.A.

GSA Books Science Editors: Marion E. Bickford and Donald I. Siegel

Library of Congress Cataloging-in-Publication Data

Preservation of random mega-scale events on Mars and Earth : influence on geologic history / edited by Mary G. Chapman, Laszlo Keszthelyi.
   p. cm. -- (Special paper ; 453)
   Includes bibliographical references.
   ISBN 978-0-8137-2453-9 (pbk.)
   1. Mars (Planet)--Geology--History. 2. Earth--History. I. Chapman, Mary G. II. Keszthelyi, Laszlo.

QB643.G46P74 2009
559.9'23—dc22

2009009036

Cover: Superposed accurately sized global views of Mars and Earth adapted from the NASA World Wind Program. The editors extend their thanks to Trent Hare (U.S. Geological Survey, Flagstaff Astrogeology Team) for the inspiration and idea for the cover artwork.

10 9 8 7 6 5 4 3 2 1

Contents

Preface v

1. *The surface of Mars: An unusual laboratory that preserves a record of catastrophic and unusual events* 1
   Mary G. Chapman

2. *Effect of impact cratering on the geologic evolution of Mars and implications for Earth* 15
   Nadine G. Barlow

3. *Megafloods and global paleoenvironmental change on Mars and Earth* 25
   Victor R. Baker

4. *Effects of megascale eruptions on Earth and Mars* 37
   Thorvaldur Thordarson, Michael Rampino, Laszlo P. Keszthelyi, and Stephen Self

5. *Terrestrial subice volcanism: Landform morphology, sequence characteristics, environmental influences, and implications for candidate Mars examples* 55
   John L. Smellie

6. *Megascale processes: Natural disasters and human behavior* 77
   Susan Werner Kieffer, Paul Barton, Ward Chesworth, Allison R. Palmer, Paul Reitan, and E-an Zen

Preface

After the Planetary Division Pardee Keynote Symposia session "Preservation of random mega-scale events on Mars and Earth: Influence on geologic history" at the GSA Seattle 2003 meeting, it was brought to our attention that the GSA usually published Special Paper publications on the topic of Pardee sessions. This GSA Special Paper's running theme between the two planets is a timely topic for a special paper, as (1) interest in Mars is running high with all the ongoing missions and (2) the science community is always put to the task of defending NASA money budgeted for planetary research. The focus of this volume is on catastrophic events that have influenced Mars and Earth and is part of the ongoing search for the correct balance between catastrophic versus uniformitarian processes. The uniformitarian paradigm guided most of terrestrial geology through the 1800s, largely ignoring the catastrophes recorded in the rock record. Uniformitarianism has since given way to an actualistic view of geologic processes that allows for the occurrence of short-duration events that cause extreme changes in the surface of the Earth. This realization was triggered, in part, by the growing recognition of the importance of these processes in extraterrestrial environments, such as the basin-forming impacts on the Moon. Many features formed by large, catastrophic, and unusual events are readily apparent on Mars, but are relatively poorly recorded on Earth. Is this difference in the record between the two planets solely the result of a distinctly lower preservation potential for large-scale random events on Earth, where gradualistic processes, such as tectonics and erosion, mask them, or is the difference predominantly due to inherent differences in gravity, atmosphere, climate, and materials between the planets? Are we biased against the recognition of stochastic megaevents on Earth by vestiges of the uniformitarian paradigm? What is the relative importance of nonuniform, catastrophic, and unusual processes, as compared to the gradualistic, uniformitarian processes, in shaping the geomorphic, climatic, and biotic history of the Earth? What can we learn about these processes on Mars that we can extend to our knowledge of the Earth? Finally, are we as humans a unique process confined to Earth that can cause unalterable terrestrial global change?

This GSA Special Paper aims to expand and update our "Geoscience Horizons" by examining various large-scale but short-duration geologic processes that have been recognized on both Earth and Mars and their preservation potentials on Earth as compared to Mars, and considers the effects of such processes on the evolution of Earth. We hope this broad spectrum of topics will be of interest to a wide range of readers.

We wish to thank all the authors who participated in the volume and were patient with the editorial process. We also thank the reviewers that improved the text.

This Special Paper is dedicated to the memory of William T. Chapman (December 21, 1919 to June 26, 2006) and David J. MacKinnon (September 23, 1944 to October 19, 2006).

Mary G. Chapman
Laszlo P. Keszthelyi

The surface of Mars: An unusual laboratory that preserves a record of catastrophic and unusual events

Mary G. Chapman*
Astrogeology Team, U.S. Geological Survey, 2255 N. Gemini Drive, Flagstaff, Arizona 86001, USA

ABSTRACT

Catastrophic and unusual events on Earth such as bolide impacts, megafloods, supereruptions, flood volcanism, and subice volcanism may have devastating effects when they occur. Although these processes have unique characteristics and form distinctive features and deposits, we have difficulties identifying them and measuring the magnitude of their effects. Our difficulties with interpreting these processes and identifying their consequences are understandable considering their infrequency on Earth, combined with the low preservation potential of their deposits in the terrestrial rock record. Although we know these events do happen, they are infrequent enough that the deposits are poorly preserved on the geologically active face of the Earth, where erosion, volcanism, and tectonism constantly change the surface. Unlike the Earth, on Mars catastrophic and unusual features are well preserved because of the slow modification of the surface. Significant precipitation has not occurred on Mars for billions of years and there appears to be no discrete crustal plates to have undergone subduction and destruction. Therefore the ancient surface of Mars preserves geologic features and deposits that result from these extraordinary events. Also, unlike the other planets, Mars is the most similar to our own, having an atmosphere, surface ice, volcanism, and evidence of once-flowing water. So although our understanding of precursors, processes, and possible biological effects of catastrophic and unusual processes is limited on Earth, some of these mysteries may be better understood through investigating the surface of Mars.

INTRODUCTION

On several occasions I have been confronted by people who wonder why we study Mars. Personal interest aside, gaining any knowledge of our universe is worthwhile because everything learned is a building block to advance our understanding; no one knows whether the outcome of planetary research will be a crucial discovery or, perhaps, lead to a development of economically valuable science spin-offs. While this explanation satisfies some people, it can also degenerate into unprofitable discourse trying to convince individuals of the value of our technological advancements and space science spin-offs. Another valid and currently popular reason to do planetary research is the need to better understand the geology of extraterrestrial worlds in order to understand the biologic evolution of life on our planet. However, the value of this necessity does not usually satisfy those who can only justify spending governmental funds when they are put toward immediately sustaining continued human prosperity and health (even when one counters that if all science had to be tied to immediate societal benefit our civilization would likely still be in the Dark Ages). The most important real and pressing need to study Mars may be because the surface of that planet is a laboratory that preserves a record of catastrophic and unusual events that have devastating effects when they occur on

*mchapman@usgs.gov

Figure 1. Examples of catastrophic and unusual events on Earth: (A) bolide impact—illustration; (B) floods—photograph of Fargo, North Dakota, in 1897; (C) volcanic eruptions—night at Mount Etna, Italy; and (D) subglacial eruption—1996 Gjálp, Vatnajökull, Iceland.

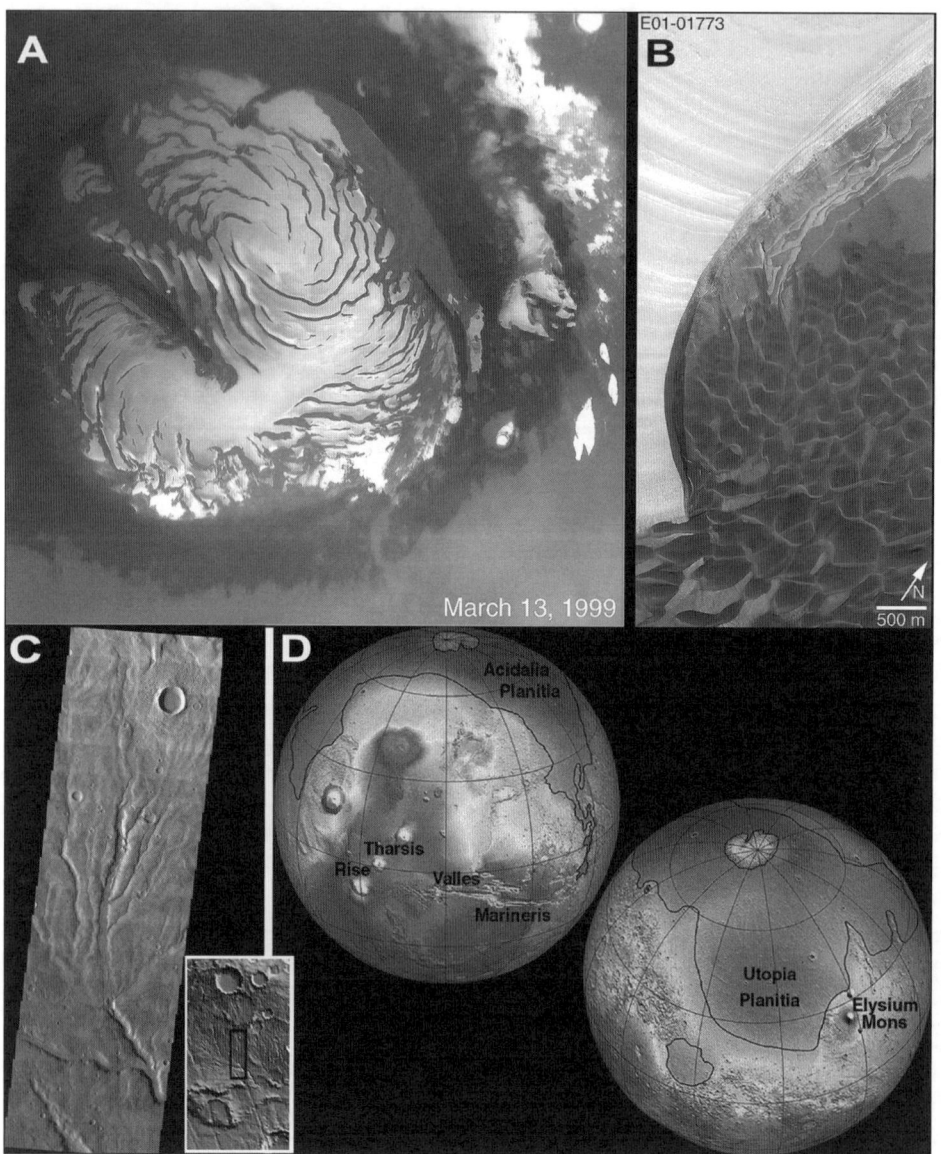

Figure 2. Evidence of modern and ancient ice on Mars. (A) Mars Global Surveyor (MGS) Mars Orbiter Camera wide angle image of Martian north polar cap (1100 km wide; MOC2–231) in early northern summer. (B) North polar ice cliff above dark sand dunes (part of MOC E01–01773; 3.3 km wide; 83.60°N, 242.02°W). (C) THEMIS VIS image showing multiple branching channels in Warrego Valles region of Mars at about lat –42.3, long 267.5 East (NASA Photojournal PIA05662; 19 m/pixel resolution; inset shows context image). (D) Hemispheric views of Mars (courtesy of Mars Orbital Laser Altimeter Team) showing feature locales and location of elevation contour –3760 ± 560 m (black line) cited by Head et al. (1999) as being consistent with the shoreline of an ancient ocean.

Earth (Fig. 1). Events such as bolide impacts (or impact cratering), megafloods, supereruptions, flood volcanism, and subice volcanism are all potential megadisasters on our planet that could cause the economic demise of countries, huge mass mortalities, and even result in mass extinctions. These megadisasters do happen, but are infrequent enough that they are not well preserved on the geologically active face of the Earth, where ongoing erosion, volcanism, and tectonism cause constant surface change. Hence, our understanding of precursors, processes, and possible biological effects of these potentially devastating megadisaster events is limited on this planet. However, we know that impact cratering, megafloods, supereruptions, flood volcanism, and subice volcanism all occurred on Mars. Therefore some of the mysteries of these megaevents may be unraveled through investigating the surface of Mars, rather than focusing on Earth.

On Mars, we can study megaevent processes on a planet that has surficial geologic processes most similar to our own. The other planets either (1) lack atmospheres or have unearthly, dense hostile atmospheres or (2) lack surface volatiles or have unearthly surfaces consisting entirely of ice. Like the Earth, Mars has an atmosphere (albeit currently a much thinner one), surface ice on both poles, and reservoirs of subsurface volatiles (Fig. 2, A and B). In addition, the planet's surface shows abundant geomorphic evidence of an ancient Earth-like atmospheric thickness that supported surface volatiles, even in equatorial areas. For example, we have known since the 1970s from Mariner 9 and Viking data that the surface has fluvial valleys and huge flood channels (Masursky, 1973; Baker and Milton, 1974).

Geological observations, such as channels, chaotically collapsed ground, and flow lobes on crater ejecta, attest to an ancient large reservoir of ground ice, near the surface or at depth, almost everywhere on Mars (Fig. 2C). These observations led Carr (1995) to suggest an ancient water inventory equivalent to a layer several hundred meters thick (if evenly distributed). Meteorite studies indicate that the crust of the planet Mars may have held two to three times as much water (Leshin, 2000). Many have suggested that water from ancient surface channels appears to have pooled in the northern lowlands, forming paleolakes or an ocean (Witbeck and Underwood, 1983; Lucchitta et al., 1986; Parker et al., 1989, 1993; Jöns, 1990; Baker et al., 1991; Scott et al., 1992; Lucchitta, 1993). Parker et al. (1989, 1993) and Jöns (1991) mapped what they interpreted to be shorelines for this ancient ocean in the northern plains of Mars. Data from the Mars Orbiter Laser Altimeter (MOLA) on the recently transmitting Mars Global Surveyor (MGS) spacecraft are consistent with these hypotheses of large standing bodies of water in the northern lowlands in the past history of Mars because a putative shoreline (mostly labeled "Contact 2" in Parker et al., 1989) closely matches an equipotential surface with a mean elevation of -3760 ± 560 m (Fig. 2d; Head et al., 1999).

Recently, MGS spacecraft Mars Orbiter Camera (MOC) images appear to indicate the presence of ground ice poleward of the 30° latitude belts (Malin and Edgett, 2000), and epithermal neutron rates of surface hydrogen abundances measured by the Neutron Spectrometer on the currently orbiting Mars Odyssey spacecraft indicate that near-surface dirty ice may exist at high latitudes and in some localized areas at lower latitudes (Feldman et al., 2004). Evidence from the Mars Exploration Rovers indicates water-alteration of Gusev site rocks and ancient aqueous environment at Meridiani Planum (Squyres et al., 2004; Herkenhoff, et al., 2004; Haskin et al., 2005). Although it has been suggested that water as ice and perhaps in liquid form may even be active in places on the Martian surface today (Malin et al., 2006), modeling based on new high-resolution images from the High Resolution Imaging Science Experiment (HiRISE) camera cast doubt on this interpretation (Pelletier et al., 2008). Albeit the planet's surface has consistently shown abundant geomorphic evidence of a preexisting Earth-like atmospheric thickness that supported surface volatiles globally, these features appear to be ancient and most formed from the release of ground water/ice to the surface and not due to erosion from precipitation. Significant precipitation has not occurred on Mars for billions of years, and there appears to be no plate tectonics, so there is little to obliterate surface features. Because these erosion processes are lacking, catastrophic and unusual megaevent features are relatively well-preserved and readily observable on spacecraft data.

Besides the dramatically better preservation states of the surface, the sizes of many geologic features on Mars are enormous relative to the Earth and therefore much easier to see and study. For example, the 300-km-diam Precambrian Vredefort impact structure in South Africa is the Earth's largest known impact structure (Fig. 3A). However, on Mars, the diameters of the Hellas, Argyre, and Utopia Planitiae impact basins are 1800, 1500, and 3800 km, respectively (Fig. 3B). This size difference is because most of Mars is much older than the Earth's surface and therefore large impact craters that date from ancient higher impact rates are observable on its surface. On Earth, surface erosion and crustal recycling have erased all huge ancient impact craters.

Catastrophic flood channels on Mars are also enormous—some extending thousands of kilometers (Fig. 4). Paleodischarges for selected Martian channels have been estimated using the standard Chezy and Manning equations appropriately modified for Mars gravity (e.g., Komar, 1979; Baker, 1982; Robinson and Tanaka, 1990; De Hon and Pani, 1993; Komatsu and Baker, 1997; Chapman et al., 2003). The results (tabulated for comparison in Komatsu and Baker, 2007) indicate that paleodischarges may have been 10–100 times greater than the largest known floods on Earth (Baker, 1982).

In addition, Martian volcanoes dwarf those on Earth: Olympus Mons, for example, is 24 km high and 500 km wide (Fig. 5). Although Mars is famous for having the Solar System's largest shield volcanoes, it also boasts immense flood lavas. Mapping suggests that much of the Martian plains are covered by flood lavas (e.g., Scott and Carr, 1978; Scott and Tanaka, 1986; Tanaka and Scott, 1987; Greeley and Guest, 1987). There are potential subice volcanoes on Mars and, while not as large as some terrestrial subaerial volcanoes, they are an order of magnitude larger than any subice edifices on Earth (Fig. 6; Chapman and Tanaka,

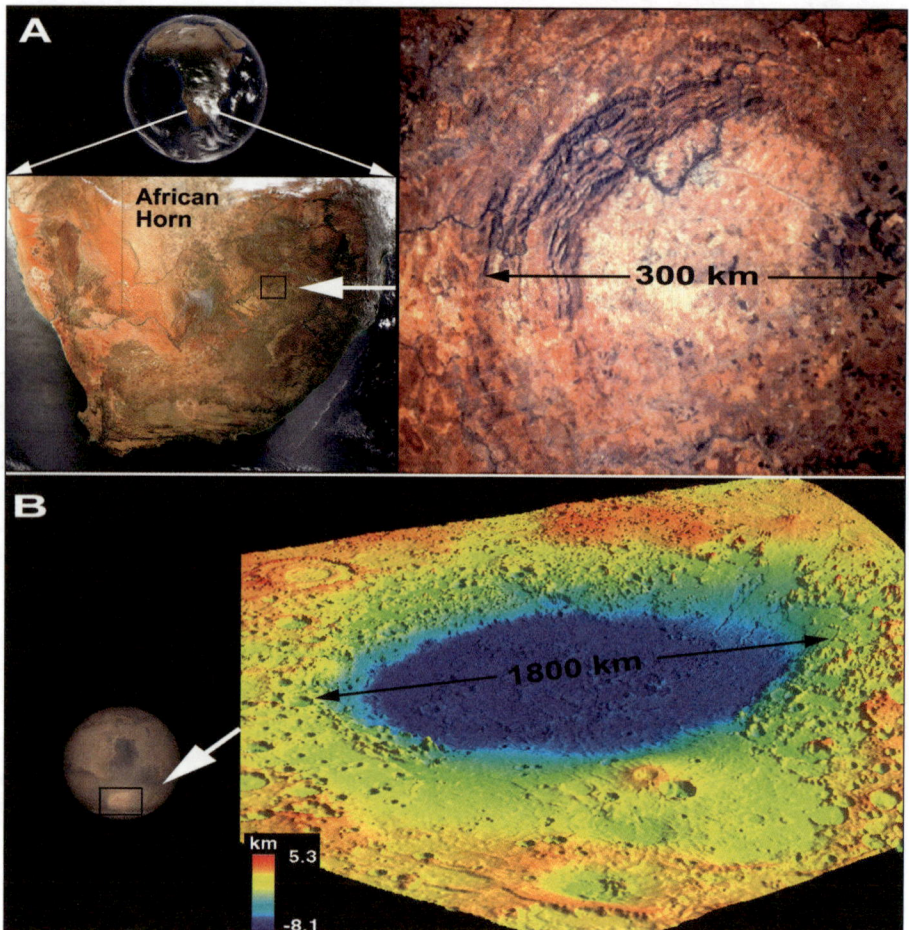

Figure 3. Comparison of large impact craters on Earth and Mars. (A) Hemisphere view of Earth showing the horn of Africa and location of Earth's largest known impact structure: the 300-km-diam Pre-Cambrian Vredefort Crater. (B) Hemisphere view of Mars showing the 1800-km-diam Hellas impact basin.

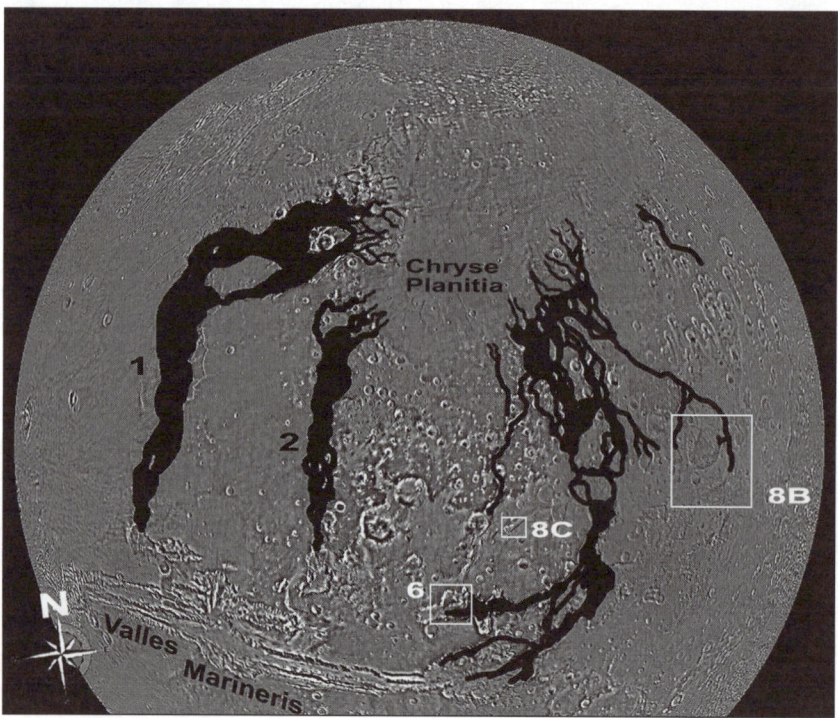

Figure 4. Most of the western hemisphere of Mars showing outflow channels on Mars (in black) that drain into the Chryse Planitia basin and the equatorial canyon of Valles Marineris; for scale, the Kasei Valles channel (1) is ~2400 km long and the Maja Valles channel (2) is ~1600 km in length; boxes show approximate locations of Figures 6, 8B, and 8C.

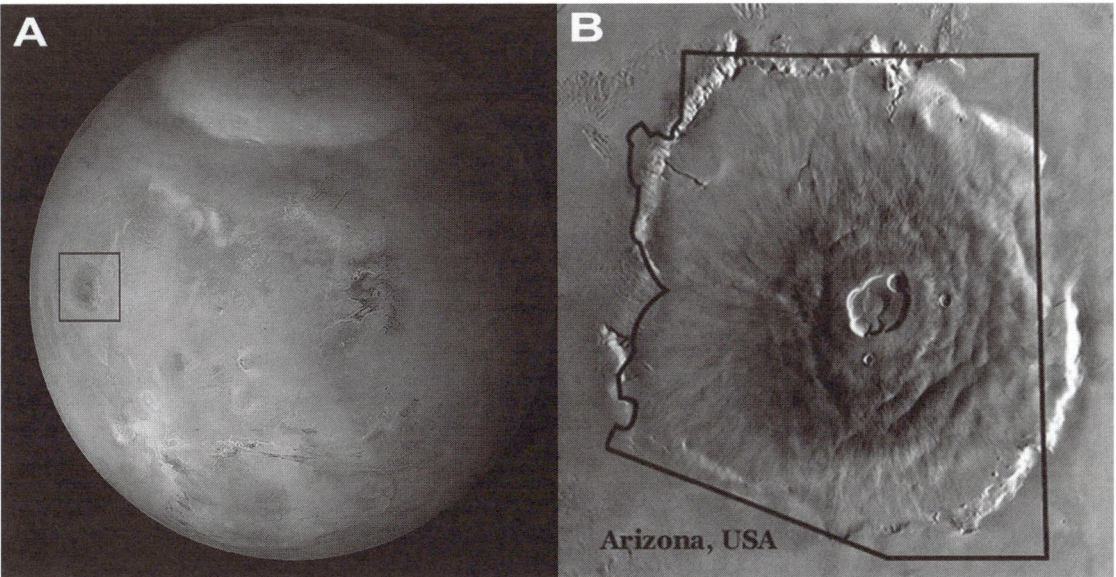

Figure 5. Images of the largest volcano in the solar system. (A) Hemisphere view of Mars showing location of Olympus Mons. (B) Olympus Mons with outline of the state of Arizona.

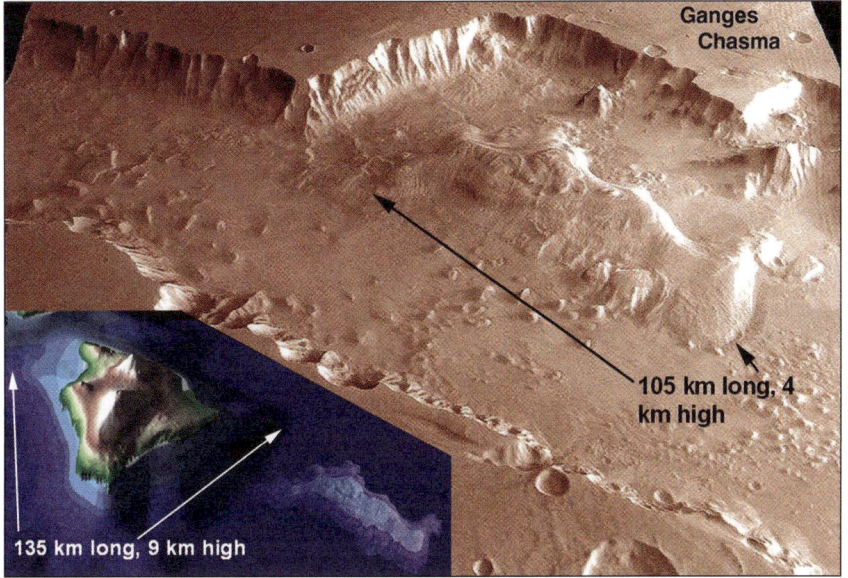

Figure 6. Perspective view of Ganges Chasma, part of Valles Marineris on Mars, and its internal mound (Viking imagery draped over MOLA topography; location shown on Fig. 4). For comparison, inset image shows topographic perspective view of the Big Island of Hawaii with the Ganges mound (note the Ganges mound would not rise above Earth's sea level).

2001; Chapman and Smellie, 2007). The reason flood lavas are more widespread and that many volcanoes grow much larger on Mars than those on Earth is because magma discharges are potentially much greater (Wilson and Head, 1994) and because there are no tectonic plates to shift and displace the volcanism.

One megaevent not known to have occurred in a larger fashion on Mars relative to the Earth are tidal waves because Mars is so cold relative to the Earth that any paleo-oceans or lakes on Mars would have likely been frozen and unable to generate catastrophic waves.

On Mars we can study megaevent processes on a planet that is most similar to our own, where the sizes of the associated geologic features are enormous and extremely well-preserved. The mysteries of catastrophic and unusual processes may be unraveled through investigating the surface of Mars. This paper, meant as an introductory chapter to the volume, discusses the evolution of our knowledge of such catastrophic and unusual processes and the role that other planets, and especially Mars, had and can have in the continued study of these processes.

One process of geologic catastrophes does not occur on Mars. This is the effect of humans in creating megacatastrophes. Human-induced megascale processes, in the past, present conditions, and future effects are discussed in this volume (see Kieffer et al., this volume). One can only hope that we can and will control human megaeffect catastrophes, so that human habitation of Mars does not become a necessity.

PLANETARY MEGAEVENTS

Studying catastrophic events on other planets in order to decipher the geologic history of the Earth is a practice that has historic precedence. Our understanding of basic geologic principles has been vastly altered due to the study of planetary geology. For example, until the later half of the 1900s, the principle of uniformitarianism was dogmatically adhered to by most geologists. This basic tenet of geologic science dictated that most of the Earth's crust formed entirely by slow, ongoing, and observable processes operating gradually over time. The original importance placed on this principle is understandable, as scientists were (and are) vested in relying on objective human observation, rather than speculation about catastrophic events that were rare or yet to be proven and observed. Eventually, however, geologic science was forced to acknowledge that catastrophic events do happen and may explain some unusual deposits. This realization was triggered, in part, by the recognition of the importance of these processes in extraterrestrial environments. In other cases, our awareness of the processes stemmed from studies of Earth, but their ramifications were only fully understood when we began to study analogous features on Mars.

Impact Cratering

The study of craters on the Moon dramatically advanced our understanding of the catastrophic nature of asteroid and comet impacts on the Earth. The scientific study of cratering began in the late 1800s, when G.K. Gilbert examined the lunar surface and first suggested the radical idea that the lunar craters were due to bolide impacts (Gilbert, 1893). Gilbert was the first Chief Geologist of the U.S. Geological Survey (USGS) and was roundly lambasted for his lunar departure from normal "terrestrial" geologic studies. Up until the Gilbert paper was published, lunar craters were considered to be volcanic and terrestrial impact craters were not recognized as such. Although a few years earlier, the large number of meteorites near Coon Butte crater in northeastern Arizona caused a mineral dealer to propose that this site was an impact crater (Foote, 1891). After presenting his paper, Gilbert was made aware of Foote's Coon Butte hypothesis. He promptly dispatched two geologists to visit the site, but was convinced from their reports that the crater was probably a volcanic maar. In 1902, mining engineer D.M. Barringer and surveyor/geologist B.C. Tilghman filed a mining claim on the Coon Butte crater and began to drill—the results of the drilling and site collection suggested that the crater was indeed formed by meteorite impact (Tilghman, 1905). Barringer (1905) renamed the feature Meteor Crater and produced a map and detailed studies supporting the impact hypothesis, as did another geologist named Merrill (1908).

None of these reports, however, changed the viewpoint of the USGS which steadfastly rejected the meteorite impact hypothesis. After spending a fortune and numerous years excavating for the bolide without finding a major metallic deposit, Barringer's hypothesis became more widely doubted. In the late 1940s, other workers were drawn to Meteor Crater and meteorite specialist H.H. Nininger collected numerous samples near the crater rim.

During this same time period (1948), Eugene (Gene) Shoemaker joined the USGS and began exploring in Colorado and Utah for uranium deposits. That first summer on his new job, thinking about nuclear fuel and rocket development during World War II, it suddenly occurred to Gene that during his lifetime man would go to the Moon. Gene was fascinated by the moon; he wanted to be the first lunar geologist and began lunar studies in his free time. Gene Shoemaker's USGS research, funded in the 1950s by the Atomic Energy Commission, consisted of mapping maars, salt anticlines, Hopi-Butte diatremes, and laccoliths of the Colorado Plateau for nuclear material research studies. Gene wanted to know where radioactive material would go in an underground volcanic explosion and so began studying the effects of shock in nuclear craters. He immediately saw the resemblance of these craters to Meteor Crater on the Colorado Plateau. In 1956 Nininger published his findings on the meteorites that he found strewn about the crater and concluded that Meteor Crater must be due to an impact, but impact mechanics were unresolved. In the period 1957–1960, Gene Shoemaker did his classic dissertation research on the structure and mechanics of meteorite impacts. In 1957, well before President Kennedy's 1961 initiation of the Moon race, Gene began to prepare a preliminary map of Copernicus Crater on the Moon. Using only photographs and telescopic measurements, he showed that he could unravel the sequence in which layers of rock were deposited. Using these lunar studies, careful comparative mapping of Meteor Crater by Shoemaker (1960) and laboratory studies of unique shock-formed minerals and their deformation features at the site (Chao et al. 1960; Kieffer, 1974) led Meteor (also known as Barringer) Crater to become the first confirmed impact crater on Earth. Baldwin (1963) went on to establish a quantitative morphology between bomb craters and lunar craters. Asteroid and comet research combined with continued study of impact craters on the Moon, and the other planets, led researchers to conclude that the Earth must have been similarly bombarded, but these terrestrial craters were either not preserved or had not yet been recognized. With this information gleaned from extraterrestrial data, workers began to actively search for and reexamine circular features on the Earth for evidence that some might indeed be impact craters. To date, 174 impact sites have been confirmed on the Earth (Earth Impact Database, 2009).

Unlike the Earth, the moon lacks an atmosphere and huge reservoirs of subsurface volatiles, which leads to better preservation of impact structures. Similarly, the cratering record is better preserved on Mars than on Earth. However, on Mars we are able to study the interactive effects between impact cratering and an atmosphere and cryosphere, without the erosional overprinting seen on Earth. As discussed elsewhere in this volume (see chapter by Barlow), our studies of Mars indicate that cratering produces heating which may have contributed volatiles to the Martian atmosphere and catalyzed some of the outflow events that carved large channels. In addition, Mars's volatile-rich environment likely contributes to the greater extent of ejecta blankets seen on this planet than on the Moon (see Barlow, this volume). The cratering record of Mars thus holds important implications for how impacts may have affected the Earth (Barlow, this volume).

Catastrophic Floods

When J. Harlen Bretz (1923a, 1923b) interpreted the Channeled Scabland of eastern Washington to be the result of a catastrophic flood of water that eroded the volcanic plateau to a scabland in perhaps only a few days, aghast geologists roundly denounced his ideas (as related by Alt, 2001). Bretz's peers, deeply enmeshed in the geologic principal of uniformitarianism, rejected his observations. In 1910, J.T. Pardee published a paper on Montana's Glacial Lake Missoula—a large body of water that existed at the edge of the vast ice sheet covering North America during the ice age from 40 ka to the last glacial maximum, ~15 ka (Pardee, 1910). Thirty years later, he interpreted the rocks bounding the east edge of the scabland to indicate that the Missoula Lake had been emptied catastrophically due to an ice dam collapse in northern Idaho, but he did not conclude that this lake had been the source of Bretz's channeled scabland floods (Pardee, 1942). The catastrophic flood hypothesis of Bretz remained largely scorned until the 1950s, when a few geologists warmed to the idea (Alt, 2001). In 1971, hydrologists from the USGS estimated the peak flood flow that drained the entire Lake Missoula at Eddy Narrows to have been at least ten times the combined flow from all the known rivers in the world (Weis and Neuman, 1971).

Fifty years after Harlen Bretz announced his Channeled Scabland hypothesis, it was one of the great surprises of modern planetary science that immense fluvial valleys were discovered on vidicon images returned by the Mariner 9 spacecraft, which went into orbit around Mars on 13 November 1971 (Masursky, 1973). The spacecraft returned over 7000 images of Mars (until it had to be shut down on 27 October 1972), which revealed huge Martian channels that extend thousands of kilometers and showed eroded geomorphologies very similar to those seen in the Channeled Scabland on Earth (Baker and Milton, 1974). These observed morphologies (such as extensive grooved channel floors) and the scales of the outflow channels indicated that they were carved by fluid flows of immense magnitude, and the similarities between the Martian outflow channels and the Channeled Scabland strongly suggested that outflow channels were carved by catastrophic water floods (Baker, 1982). Estimates indicate that the paleodischarges on the Martian channels may have been 10–100 times greater than the largest known floods on Earth, such as the enormous Missoula glacial floods and those recorded in deposits along the Mississippi River (Baker, 1982).

On Earth, catastrophic floods are formed only by outbursts of surface water dammed behind ice (the Channeled Scabland), subglacial ice outbursts (Icelandic jökulhlaups), and spillovers of huge lakes (ancient Lake Bonneville flood). The Martian floods may have had similar triggers, but must have been sourced by much larger volumes of water (Komatsu and Baker, 2007). Driven by studies of the Martian channels, there have been recent discoveries of catastrophic flood features in northern Eurasia (e.g., Baker et al., 1993; Komatsu et al., 1997); making it increasingly clear that catastrophic floods occurred repeatedly on Earth during the ice age (Komatsu and Baker, 2007). Therefore the ancient surface of Mars and its water endowment that produced catastrophic flood channels (Baker et al., 1991) may have been similar to Earth's surface during its ice ages. The limited erosion on Mars has left the catastrophic channels formed on its surface well-preserved. In contrast on Earth, water, glacial ice, wind erosion, volcanic deposition, sea-level rise, urban growth, and vegetation cover have altered the surface tremendously since the last ice age. Therefore the study of outflow channels on Mars helps us to understand the precursors, processes, and possible effects of catastrophic floods on Earth (see Baker, this volume).

Megaeruptions

Extremely large volcanic eruptions or megaeruptions occur when giant bodies of molten magma rise up through Earth's crust to erupt onto the surface. Depending on whether the magma bodies are felsic or mafic, potentially devastating effects from these megaplumes of molten magma are (1) huge volcano-tectonic collapse structures (supercalderas) and widespread ash deposits or (2) basaltic flood lavas (see Thordarson et al., this volume).

Supercaldera Eruptions

Only a few supercalderas have been identified on the active surface of the Earth. All of these are classified as resurgent calderas, meaning that they have erupted several times at somewhat regular intervals. The Toba caldera produced the largest global eruption in the last 1 m.y. of the Earth's history. This caldera is 30 × 100 km (Fig. 7) and has a total relief of 5100 feet (1700 m). The caldera probably formed in stages. Large eruptions occurred 840,000, ~700,000, and 75,000 yr ago (Chesner et al., 1991; Rampino and Self, 1992, 1993). Based on DNA studies, some workers have suggested that a genetic bottleneck occurred around 70,000–75,000 yr ago when our population was reduced to almost endangered-species levels. This was suggested to be attributed to the Toba eruption having caused 1000 years of the coldest temperatures during the Later Pleistocene (~71–70 ka; Ambrose, 1998). The prolonged volcanic winter combined with immediate eruption killing effects (such as blast fronts, tidal waves, fires, ash blanket vegetation kills, and animal suffocation, and later ash-inhaled respiratory disease) in the tropical refugia hundreds of kilometers around Toba may have decimated ancient human populations (see Thordarson et al., this volume). Although archaeological evidence recently discovered in India does not appear to support such a catastrophic effect on modern humans (rather than Neanderthals or other hominines) from the Toba eruption (Petraglia et al., 2007); supercaldera eruptions still retain a high potential for human calamity. One of the largest calderas on the Earth is the active Yellowstone Caldera, which had huge prehistoric eruptions 2.0, 1.3, and 0.6 m.y. ago. Each eruption formed a caldera and extensive layers of thick proximal pyroclastic-flow deposits. The eruptions also produced thick distal ash-fall deposits over the entire western and central United States. The youngest of the nested Yellowstone calderas is an elliptical depression or basin nearly 50 km long and 80 km wide that occupies much of Yellowstone National Park. Latent subsurface heat from this caldera powers the park's famous geysers, hot springs, fumaroles, and mud pots. Compacted ash thicknesses of 0.2 m from this event occur ~1500 km from the source (Wood and Kienle, 1990; Sarna-Wojcicki and Meyer, 1984). Ash falls of this magnitude would have disastrous effects on the ecosystems within this radius. As a species, we are at risk from the inevitable reoccurrence of resurgent supercaldera eruptions, but we are again limited in terrestrial study of these features on the active surface of the Earth where our knowledge of their full size

Figure 7. LandSat image of the Toba Caldera in Indonesia; inset illustrates geographic detail.

and measure is compromised by ongoing erosion and plate tectonic destruction. However, megacalderas on Mars dwarf even the largest of those observed on Earth. For example, the 60 × 80-km-diam caldera complex of Olympus Mons contains six major collapse structures from resurgent events (Fig. 5), the large caldera of Pavonis Mons is roughly circular with a diameter of 80–100 km, the Arsia Mons caldera is a 110-km-diam caldera, Uranius Patera is a 75 × 150-km-diam caldera, and Alba Patera is roughly circular with a 150-km-diam caldera (Mouginis-Mark et al., 2007). In addition, it has been suggested that areas of chaotic materials within circular depressions, like Aram Chaos, are perhaps also volcano-tectonic calderas (Fig. 7; Chapman, 2002). This idea may not be too farfetched, as MOLA-derived topography of Aram Chaos indicates that like the Valles Caldera in New Mexico, Aram Chaos is on a regional topographic high and contains interior mounds positioned asymmetric to its center (Fig. 7; Chapman, 2002), and like terrestrial calderas MOC images of Aram show possible small internal volcanic edifices (Lanz and Jaumann, 2001). Even larger Martian calderas may exist. Perhaps the elliptically shaped chasms of Valles Marineris are very large explosive calderas that formed in association with ground ice (Fig. 8; Chapman and Smellie, 2007). If so, these chasmata are huge, ranging from 300 to 800 km long, 100 to 300 km wide, and up to 8 km deep. Our understanding of the surface effects of supercaldera eruptions can be improved by investigating high-resolution imagery of the megastructures on Mars. Although supercalderas may be more accessible on Earth, their size and easily filled depression form makes them hard to identify; in addition, erosion and plant growth have masked the evidence of surface damage from megaeruptions. A final comment is that no supercaldera events are known to have interacted with ice or formed beneath ice sheets on Earth, although these have been postulated for Mars. A supereruption occurring beneath ice in Earth's polar areas might have devastating consequences for the global ice inventory and instigate sea-level rise. Modeling suggests that a large silicic eruption with a magma discharge rate of 10^9 kg s^{-1} would melt its way through a 3000-m-thick ice sheet in hours or days (Gudmundsson, 2003). A GIS analysis of global inundation impacts from sea-level rise suggests that a relatively small 1 m rise in global sea level would inundate over a 10×10^6 km^2 area, affecting 108 million people worldwide with displacement and loss of agricultural and cultural resources (Rowley et al., 2007).

Flood Lava Eruptions

Flood lavas form vast lava plains without major volcanic edifices, and their eruptions can have significant and global effects on planets (Self et al., 2005; Keszthelyi and McEwen, 2007; Thordarson et al., this volume). Of greatest importance to human civilization is the apparent temporal coincidence of flood basalt eruptions with mass extinctions, in that all major extinction events in the past 300 Ma have flood basalt eruptions occurring at the same time (Rampino and Stothers, 1988; Haggerty, 1996; Wignall, 2001). Although the precise role that flood lava eruptions play in mass extinctions is still highly uncertain, it may

Figure 8. Oblique views of (A) a relief map of the Jemez Mountains, New Mexico, showing the resurgent Valles Caldera (illumination from upper left; R. Bailey, U.S. Geological Survey), (B) Aram Chaos (illumination from right; courtesy of the MOLA Team), and (C) Hrad Vallis, part of Valles Marineris; north is toward the right; location of B and C shown on Figure 4.

be due to effects of released gases on the environment and global climate (Thordarson et al., this volume).

Flood-basalt provinces are scattered across the Earth, but make up only a few percent of its surface area (Keszthelyi and McEwen, 2007). Less than a dozen well-preserved and well-exposed continental flood-basalt provinces are known with traces of many older flood basalts largely erased by the vigorous geologic activity on Earth; most of the largest and best-preserved flood-basalt provinces (e.g., the Ontong-Java and Kerguelen Plateaus) are found on the ocean floor, limiting access to them (Keszthelyi and McEwen, 2007). In contrast, flood lavas make up a significant part of the surface of Mars. A very conservative estimate by Greeley and Schneid (1991) states that at least 46% of Mars is covered by volcanics and 82.4% of this area is composed of volcanic plains and plateaus; therefore a minimum of 38% of Mars's surface is covered by flood lavas (Keszthelyi and McEwen, 2007). The study of flood basalts on Mars may help us to improve our understanding of these eruptions on both Earth and Mars. The slow rate of erosion on Mars provides a unique view of the upper surfaces of megascale flood-lava flows that can be obscured or eroded on Earth. Studies of these surfaces can increase our knowledge of the rheological properties of flood basalts.

Subice Volcanism

Terrestrial subice eruptions and some ancient megafloods on Earth form in volcanically active areas near transitory ice sheets of polar and subpolar regions. When volcanoes erupt underneath ice, they melt large volumes of water, creating a vault and meltwater lake within which tephra piles are constructed (Björnsson, 1988; Smellie, 2000). A characteristic feature of volcanic eruptions within ice is the typical release of the meltwater through voluminous floods or jökulhlaups (Björnsson, 1988, 1992; Tweed and Russell, 1999; Gudmundsson et al., 2004). The floods can also mobilize very large quantities of sediment to form hyperconcentrated floods and lahars, the effects of which can be catastrophic on human lives and a local economy (Major and Newhall, 1989). Meltwater hydraulics during subglacial eruptions is important for understanding, predicting, and mitigating the impact of local meltwater release on vulnerable human communities (see Smellie, this volume).

Historically, Icelanders have suffered the most from these types of floods, as subglacial eruptions are frequent in Iceland, with the majority occurring within the Vatnajökull ice cap. Owing to low activity in recent decades, the first such eruption to be monitored in any detail was the 13-day-long fissure eruption in Gjálp, Vatnajökull in October 1996 (Gudmundsson et al., 1997). Although the jökulhlaup from this eruption caused major damage to highways and farms on the southeast of Vatnajökull (Snorrason et al., 2002), it pales in comparison to huge areas of jökulhlaup flood materials deposited between 2500 and 2000 yr ago northwest of Vatnajökull down Jökulsá á Fjöllum (Waitt, 2002). A repeat of this type of flood down Jökulsá á Fjöllum, Iceland's largest glacial river and a now significantly populated area, would cause a huge disaster in Iceland.

In addition to localized Icelandic disasters, there are also places where future volcano-ice eruptions could have global catastrophic affects. These areas are beneath the West Antarctic ice sheet where there are active but dormant subice volcanoes (LeMasurier, 1990; Blankenship et al., 1993; Behrendt et al., 2002, 2004; Corr and Vaughan, 2008). Antarctica contains more than 90% of the world's ice, and the massive west Antarctic ice sheet, previously assumed to be stable, is starting to collapse due to climate change (NewScientist, 2005). The UN Intergovernmental Panel on Climate Change 2006 Fourth Assessment report notes that our climate is warming while our carbon dioxide (greenhouse gas) emissions have risen during the past five years by 3% (well above the 0.4% a year average of the previous two decades) and predicts a sea-level rise of 17 in. by 2100 (The Sunday Telegraph, 2006). These sea-level rise predictions are based on climate change alone and do not take into account the added consequences of a possible subice eruption beneath the West Antarctic ice sheet. A large subice eruption could rapidly instigate greater West Antarctic ice sheet collapse—adding to the devastating consequences of global sea-level rise. This potentially catastrophic event would be a megadisaster.

By definition active terrestrial subice volcanoes are buried beneath ice sheets and therefore are difficult to study, but huge ancient subice volcanoes also appear to have existed on Mars. As discussed in the Introduction, numerous lines of evidence indicate that ancient Mars had an Earth-like atmospheric thickness that supported surface volatiles. Colder temperatures on Mars would have rapidly frozen ponded areas of surface water. Frozen waters likely interacted with volcanoes, in view of crater density data and extensive morphological evidence that suggest volcanism and surface ice/water occurred widely throughout Mars's geologic history (Tanaka et al., 1992, Chapman et al., 2000). Indeed, the interaction of magma, groundwater, and ice may have been critical in the evolution of Mars (Head, 2007), and some Martian geomorphic features have been interpreted to be analogous to terrestrial subice volcanoes or tuyas and hyaloclastic ridges (Hodges and Moore, 1978; Allen, 1979; Anderson, 1992; Chapman, 1994, 2003; Chapman and Tanaka, 2001, 2002; Ghatan and Head, 2002). Besides Earth, Mars is the only planet we know that may have subice volcanoes.

Possible edifices have been identified in the planet's northern plains and near the equator in Valles Marineris, a canyon that is 4000 km long and from 2 to 8 km deep–a length equal to the entire United States (Fig. 4). The canyon system consists of isolated and interconnected elliptical and linear troughs or chasmata near the equator of Mars, and mounds within the troughs are suggested to be subice volcanoes (Croft, 1990; Lucchitta et al., 1994; Chapman and Tanaka, 2001; Komatsu et al., 2004). These mounds are huge features that are an order of magnitude larger than subice volcanoes we see on Earth (Fig. 6; Chapman and Smellie, 2007). These subice features could be so large relative to Earth because magma discharges on Mars are potentially much greater (Wilson and Head, 1994), there are no tectonic plates on that planet to shift and displace the volcanism, and the thickness of ice that has been suggested to fill the great void of Valles Marineris exceeds the thickness of any ice sheet on Earth. However, this scenario may have been radically different on a pre-Cambrian snowball Earth (Kopp et al., 2005). Because the mounds in Valles Marineris are smaller than the Big Island of Hawaii, it is possible that subice volcanoes erupted beneath the frozen oceans of ancient Earth to form rather large edifices with volumes comparable to Hawaiian volcanoes (Fig. 6).

CONCLUSION

In stark contrast to Mars, on Earth the continual process of surface modification by erosion and deposition in water-filled basins dominates the easily accessible continental rock record. The oceanic rock record contains more volcanic material, but is fairly inaccessible and subject to destruction at tectonic plate margins. Rare catastrophic events are poorly preserved on the geologically active face of the Earth, where erosion, volcanism, and tectonism constantly change the surface making them difficult to study.

However, the ancient surface of Mars preserves geologic features and deposits formed by catastrophic and unusual events. Many features formed by large and/or punctuated events are readily apparent on the surface of Mars but are relatively poorly recorded on Earth where gradual processes such as tectonics and erosion erase them. Studying such features on Mars may help clarify the relative importance of random, catastrophic processes, as compared to gradual uniform processes, in shaping the geomorphic, climatic, and biotic history of the Earth. Catastrophic and unusual events can have devastating effects when they occur on our planet that could cause the economic demise of countries, huge mass mortalities, and even result in mass extinctions of all life. Study of the almost perfectly preserved features that result from bolide impacts, megafloods, supereruptions, flood volcanism, and subice volcanism on Mars may help us to understand their precursors, processes, and possible effects on Earth.

On a final note, although this chapter discusses and supports the suggestion that an important answer to "why do we study Mars?" is that a better understanding of catastrophic and unusual events on Earth may be achieved by studying Mars, there is another valid (albeit less scientific) suggestion. In the present digital age, many first-world citizens have evolved into highly isolated individuals who work and live lifestyles devoid of real connection to the natural world that surrounds them. Humans innately feel the loss of this connection and many respond with depression and loneliness. Many fill the void by focusing on time-consuming activities, aggressively pursuing income/fame, practicing belief systems that promote connections, or turning the focus outward onto human interactions and family. However, studying any aspect of our world, the Sun, other planets of our solar system (like Mars), and the universe, can allow a connection to all that surrounds us. We are part of a collective whole that consists of not only everything on the Earth, but all within the cosmos. Learning about any aspect of the surrounding natural world, universe, and cosmos helps to define us and our place in nature—grounding us to reality. Some religions even define this collective whole as God. In conclusion, a real and valid reason to study Mars (or the aerodynamics of grasshopper wings, the physics of tornadoes, evolutionary trends of dinosaur species, and the decay rate of stars billions of light years away, and any other research topic ad infinitum) is because humans crave and want to better know their place in nature. Although counter to some parochial cultural beliefs, my subjective feeling is that scientific research allows us to better know and appreciate the wonder of it all.

ACKNOWLEDGMENTS

The author thanks Lazlo Keszthelyi and Windy Jaeger for careful reviews of the paper, the contributing authors of this volume, and the Geological Society of America for the initial Pardee Symposium.

REFERENCES CITED

Allen, C.C., 1979, Volcano-ice interactions on Mars: Journal of Geophysical Research, v. 84, p. 8048–8059, doi: 10.1029/JB084iB14p08048.

Alt, D., 2001, Glacial Lake Missoula and its humongous floods: Mountain Press Publishing, Missoula, Montana, 197 p.

Ambrose, S.H., 1998, Late Pleistocene human population bottlenecks, volcanic winter, and differentiation of modern humans: Journal of Human Evolution, v. 34, no. 6, p. 623–651, doi: 10.1006/jhev.1998.0219.

Anderson, D.M., 1992, Glaciation in Elysium: MSATT Workshop Polar Regions of Mars (Abstract), Lunar and Planetary Science Institute Technical Report 92–08, p. 1.

Baker, V.R., 1982. The Channels of Mars: University of Texas Press, Austin, Texas, 198 p.

Baker, V.R., 2009, this volume, Megafloods and global paleoenvironmental change on Mars and Earth, in Chapman, M.G., and Keszthelyi, L.P., eds., Preservation of random megascale events on Mars and Earth: Influence on geologic history: Geological Society of America Special Paper 453, doi: 10.1130/2009.453(03).

Baker, V.R., and Milton, D.J., 1974, Erosion by catastrophic floods on Mars and Earth: Icarus, v. 23, p. 27–41, doi: 10.1016/0019-1035(74)90101-8.

Baker, V.R., Strom, R.G., Gulick, V.C., Kargel, J.S., Komatsu, G., and Kale, V.S., 1991, Ancient ice sheets and the hydrological cycle on Mars: Nature, v. 352, p. 589–594, doi: 10.1038/352589a0.

Baker, V.R., Benito, G., and Rudoy, A.N., 1993, Paleohydrology of Late Pleistocene superflooding, Altai Mountains, Siberia: Science, v. 259, p. 348–350, doi: 10.1126/science.259.5093.348.

Baldwin, R.B., 1963, The Measure of the Moon: Chicago University, Chicago Press, 88 p.

Barlow, N.G., 2009, this volume, Effect of impact cratering on the geologic evolution of Mars and implications for Earth, in Chapman, M.G., and Keszthelyi, L.P., eds., Preservation of random megascale events on Mars and Earth: Influence on geologic history: Geological Society of America Special Paper 453, doi: 10.1130/2009.453(02).

Barringer, D.M., 1905, Coon Mountain and its crater: Proceedings of the Academy of National Sciences, Philadelphia, v. LVII, p. 861–886.

Behrendt, J.C., Blankenship, D.D., Morse, D.L., Finn, C.A., and Bell, R.E., 2002, Subglacial volcanic features beneath the West Antarctic Ice Sheet interpreted from aeromagnetic and radar ice sounding, in Smellie, J.L., and Chapman, M.G., eds., Volcano–Ice Interactions on Earth and Mars: Geological Society, London, Special Publications, v. 202, p. 337–355.

Behrendt, J.C., Blankenship, D.D., Morse, D.L., and Bell, R.E., 2004, Shallow-source aeromagnetic anomalies observed over the West Antarctic Ice Sheet compared with coincident bed topography from radar ice sounding—new evidence for glacial "removal" of subglacially erupted late Cenozoic rift-related volcanic edifices: Global and Planetary Change, v. 42, p. 177–193, doi: 10.1016/j.gloplacha.2003.10.006.

Björnsson, H., 1988, Hydrology of ice caps in volcanic regions: Vísindafélag Íslendinga: Societas Scientarium Islandica, v. 45, p. 1–139.

Björnsson, H., 1992, Jökulhlaups in Iceland: Prediction, characteristics and simulation: Annals of Glaciology, v. 16, p. 95–106.

Blankenship, D.D., Bell, R.E., Hodge, S.M., Brozena, J.M., Behrendt, J.C., and Finn, C.A., 1993, Active volcanism beneath the West Antarctic ice sheet: Nature, v. 361, p. 526–529, doi: 10.1038/361526a0.

Bretz, J.H., 1923a, Glacial drainage on the Columbia Plateau: Geological Society of America Bulletin, v. 34, p. 573–608.

Bretz, J.H., 1923b, The Channeled Scabland of the Columbia Plateau: The Journal of Geology, v. 31, p. 617–649.

Carr, M.H., 1995, The martian drainage system and the origin of networks and fretted channels: Journal of Geophysical Research, v. 100, p. 7479–7507, doi: 10.1029/95JE00260.

Chao, E.C.T., Shoemaker, E.M., and Madsen, B.M., 1960, First natural occurrence of coesite from Meteor Crater, Arizona: Science, v. 132, no. 3421, p. 220–222, doi: 10.1126/science.132.3421.220.

Chapman, M.G., 1994, Evidence, age, and thickness of a frozen paleolake in Utopia Planitia, Mars: Icarus, v. 109, p. 393–406, doi: 10.1006/icar.1994.1102.

Chapman, M.G., 2002, Layered, massive, and thin sediments on Mars: Possible Late Noachian to Late Amazonian tephra?, in Smellie, J.L., and Chapman, M.G., eds., Volcano–Ice Interactions on Earth and Mars: Geological Society, London, Special Publications, v. 202, p. 273–293.

Chapman, M.G., 2003, Sub-ice volcanoes and ancient oceans/lakes: A Martian challenge, in Subglacial lakes' detection, outbursts mechanisms and consequences: Global and Planetary Change, v. 35, p. 185–198, doi: 10.1016/S0921-8181(02)00126-1.

Chapman, M.G., and Smellie, J.L., 2007, Mars interior layered deposits and terrestrial sub-ice volcanoes compared: Observations and interpretations of similar geomorphic characteristics, in Chapman, M.G., ed., The Geology of Mars: Evidence from Earth-based Analogs: Cambridge, UK, Cambridge University Press, p. 178–210.

Chapman, M.G., and Tanaka, K.L., 2001, Interior trough deposits on Mars: Sub-ice volcanoes?: Journal of Geophysical Research, v. 106, no. E5, p. 10,087–10,100, doi: 10.1029/2000JE001303.

Chapman, M.G., and Tanaka, K.L., 2002, Related magma-ice interactions: Possible origin for chasmata, chaos, and surface materials in Xanthe, Margaritifer, and Meridiani Terrae, Mars: Icarus, v. 155, no. 2, p. 324–339, doi: 10.1006/icar.2001.6735.

Chapman, M.G., Allen, C.C., Gudmundsson, M.T., Gulick, V.C., Jakobsson, S.P., Lucchitta, B.K., Skilling, I.P., and Waitt, R.B., 2000, Volcanism and ice interactions on Earth and Mars, in Gregg T.K.P., and Zimbelman J.R., eds., Deep Oceans to Deep Space: Environmental Effects on Volcanic Eruptions: New York, Kluwer Academic/Plenum Publishers, p. 39–74.

Chapman, M.G., Gudmundsson, M.T., Russell, A.J., and Hare, T.M., 2003, Possible Juventae Chasma sub-ice volcanic eruptions and Maja Valles ice outburst floods, Mars: Implications of MGS crater densities, geomorphology, and topography: Journal of Geophysical Research, v. 108, no. E10, doi: 10.1029/2002JE002009.

Chesner, C.A., Rose, W.I., Deino, A., Drake, R., and Westgate, J.A., 1991, Eruptive history of the Earth's largest Quaternary caldera (Toba, Indonesia) clarified: Geology, v. 19, p. 200–203, doi: 10.1130/0091-7613 (1991)019<0200:EHOESL>2.3.CO;2.

Corr, H.F.J., and Vaughan, D.G., 2008, A recent volcanic eruption beneath the West Antarctic ice sheet: Nature Geoscience, v. 1, p. 122–125, doi: 10.1038/ngeo106.

Croft, S.K., 1990, Geologic map of the Hebes Chasma quadrangle, V. M., 500K, 00077 (abs.): National Aeronautics and Space Administration Technical Memorandum, v. 4210, p. 539–541.

De Hon, R.A., and Pani, E.A., 1993, Duration and rates of discharge: Maja Valles, Mars: Journal of Geophysical Research, v. 98, p. 9129–9138, doi: 10.1029/93JE00535.

Earth Impact Database, 2009, Website: http://www.unb.ca/passc/ImpactDatabase/, Planetary and Space Science Centre, Department of Geology, University of New Brunswick, Canada.

Feldman, W.C., Pettyman, T.H., Maurice, S., Plant, J.J., Bish, D.L., Vaniman, D.T., Mellon, M.T., Metzger, A.E., Squyers, S.W., Karunatillake, S., Boynton, W.V., Elphic, R.C., Funsten, H.O., Lawrence, D.J., and Tokar, R.L., 2004, Global distribution of near-surface hydrogen on Mars: Journal of Geophysical Research, v. 109, no. E09006, doi: 10.1029/2003je002160.

Foote, A.E., 1891, A new locality for meteoric iron with a preliminary notice of the discovery of diamonds in the iron: Proceedings of the American Association for the Advancement of Science, v. XL, p. 279–283.

Ghatan, G., and Head, J.W., 2002, Candidate subglacial volcanoes in the south polar region of Mars: Morphology, morphometry, and eruption condition: Journal of Geophysical Research, v. 107, no. E7, doi: 10.1029/2001JE001519.

Gilbert, G.K., 1893, The Moon's face: Bulletin of the Philosophical Society of Washington, v. XII, p. 241–292.

Greeley, R., and Guest, J.E., 1987, Geologic map of the eastern equatorial region of Mars: United States Geological Survey Miscellaneous Investigation Series Map I-1802-B, 1:15,000,000 scale.

Greeley, R., and Schneid, B.D., 1991, Magma generation on Mars: Amounts, rates, and comparisons with Earth, Moon, and Venus: Science, v. 254, p. 996–998, doi: 10.1126/science.254.5034.996.

Gudmundsson, M.T., 2003, Large-scale ice-volcano interaction: Possible effects of large silicic caldera-forming eruptions and superplume events under ice sheets: Geological Society of America Abstracts with Programs, v. 35, no. 6, p. 432.

Gudmundsson, M.T., Sigmundsson, F., and Björnsson, H., 1997, Ice-volcano interaction of the 1996 Gjálp subglacial eruption, Vatnajökull, Iceland: Nature, v. 389, p. 954–957, doi: 10.1038/40122.

Gudmundsson, M.T., Sigmundsson, F., Björnsson, H., and Högnadóttir, T., 2004, The 1996 eruption at Gjalp, Vatnajokull ice cap, Iceland: Efficiency of heat transfer, ice deformation and subglacial water pressure: Bulletin of Volcanology, v. 66, p. 46–65, doi: 10.1007/s00445-003-0295-9.

Haggerty, B.M., 1996, Episodes of flood-basalt volcanism defined by $^{40}Ar/^{39}Ar$ age distributions: Correlation with mass extinctions?: Journal of Undergraduate Science, v. 3, p. 155–164.

Haskin, L.A., Wang, A., Jolliff, B.L., McSween, H.Y., Clark, B.C., Des Marais, D.J., McLennan, S.M., Tosca, N.J., Hurowitz, J.A., Farmer, J.D., Yen, A.S., Squyres, S.W., Arvidson, R.E., Klingelhoefer, G., Schroeder, C., de Souza, P.A., Jr., Ming, D.W., Gellert, R., Zipfel, J., Brueckner, J., Bell, J.F., III, Herkenhoff, K.E., Christensen, P.R., Ruff, S., Blaney, D., Gorevan, S., Cabrol, N.A., Crumpler, L.S., Grant, J., and Soderblom, L., 2005, Water alteration of rocks and soils on Mars at the Spirit rover site in Gusev crater: Nature, v. 436, no. 7047, p. 66–69, doi: 10.1038/nature03640.

Head, J.W., 2007, The geology of Mars: New insights and outstanding questions, in Chapman, M.G., ed., The Geology of Mars: Evidence from Earth-based Analogs: Cambridge, UK, Cambridge University Press, p. 1–46.

Head, J.W., III, Hiesinger, H., Ivanov, M.A., Kreslavsky, M.A., Pratt, S., and Thomson, B.J., 1999, Possible ancient oceans on Mars: Evidence from Mars Orbiter Laser Altimeter data: Science, v. 286, p. 2134–2137, doi: 10.1126/science.286.5447.2134.

Herkenhoff, K.E., Squyres, S.W., Arvidson, R.E., Bass, D.S., Bell, J.F., III, Bertelsen, P., Ehlmann, B.L., Farrand, W.H., Gaddis, L., Greeley, R., Grotzinger, J.P., Hayes, A.G., Hviid, S.F., Johnson, J.R., Jolliff, B., Kinch, K.M., Knoll, A.H., Madsen, M.B., Maki, J.N., McLennan, S.M., McSween, H.Y., Ming, D.W., Rice, J.W., Richter, L., Sims, M., Smith, P.H., Soderblom, L.A., Spanovich, N., Sullivan, R.J., Thompson, S., Wdowiak, T., Weitz, C.M., and Whelley, P., 2004, Evidence from Opportunity's Microscopic Imager for water on Meridiani Planum: Science, v. 306, no. 5702, p. 1727–1730, doi: 10.1126/science.1105286.

Hodges, C.A., and Moore, H.J., 1978, Tablemountains of Mars: Abstracts of Papers Submitted to the 9th Lunar and Planetary Science Conference, Lunar and Planetary Institute, Houston, v. IX, p. 523–525.

Jöns, H.P., 1990, Das relief des Mars: Versuch einer zusammenfassenden ubersicht: Geologische Rundschau, v. 79, p. 131–164, doi: 10.1007/BF01830452.

Jöns, H.P., 1991, Interpretative analysis of the relief of the surface of Mars, Lithographisches Institut, Berlin, scale 1:30,000,000.

Keszthelyi, L., and McEwen, A., 2007, Comparison of flood lavas on Earth and Mars, in Chapman, M.G., ed., The Geology of Mars: Evidence from Earth-based Analogs: Cambridge, UK, Cambridge University Press, p. 126–150.

Kieffer, S.W., 1974, Shock metamorphism of the Coconino Sandstone at Meteor Crater: Meteoritical Society Annual Meeting August 7, 1974, Guidebook 37, Guidebook to the geology of Meteor Crater, Arizona, p.12–19.

Kieffer, S.W., Barton, P., Chesworth, W., Palmer, A.R., Reitan, P., and Zen, E., 2009, this volume, Megascale processes: Natural disasters and human behavior, in Chapman, M.G., and Keszthelyi, L.P., eds., Preservation of random megascale events on Mars and Earth: Influence on geologic history: Geological Society of America Special Paper 453, doi: 10.1130/2009.453(06).

Komar, P.D., 1979, Comparison of the hydraulics of water flows in Martian outflow channels with flows of similar scale on Earth: Icarus, v. 37, p. 156–181, doi: 10.1016/0019-1035(79)90123-4.

Komatsu, G., and Baker, V.R., 1997, Paleohydrology and flood geomorphology of a Martian outflow channel: Ares Vallis: Journal of Geophysical Research, v. 102, p. 4151–4160, doi: 10.1029/96JE02564.

Komatsu, G., and Baker, V.R., 2007, Formation of valleys and cataclysmic flood channels on Earth and Mars, in Chapman, M.G., ed., The Geology of Mars: Evidence from Earth-based Analogs: Cambridge, UK, Cambridge University Press, p. 297–321.

Komatsu, G., Clute, S.K., and Baker, V.R., 1997, Preliminary remote-sensing assessment of Pleistocene cataclysmic floods in central Asia: Abstracts of Papers Submitted to the 28th Lunar and Planetary Science Conference, Lunar and Planetary Institute, Houston, v. XXVIII, p. 747–748.

Komatsu, G., Ori, G., Ciarcelluti, P., and Yuri, Y.D., 2004, Interior layered deposits of Valles Marineris, Mars: Analogous subice volcanism related to Baikal Rifting, Southern Siberia: Planetary and Space Science, v. 52, p. 167–187, doi: 10.1016/j.pss.2003.08.003.

Kopp, R.E., Kirschvink, J.L., Hilburn, I.A., and Nash, C.Z., 2005, The Paleoproterozoic snowball Earth: A climate disaster triggered by the evolution of oxygenic photosynthesis: Proceedings of the National Academy of Sciences of the United States of America, v. 102, no. 32, p. 11,131–11,136, doi: 10.1073/pnas.0504878102.

Lanz, J.K., and Jaumann, R., 2001, Possible volcanic constructs in Aram Chaos revealed by MOC and their impact on outflow channel genesis (abs.): 32nd Lunar and Planetary Science Conference, Lunar and Planetary Institute, March 15–18, Houston, Texas, LPSC CD 32, abstract #1574.

LeMasurier, W.E., 1990, Late Cenozoic volcanism on the Antarctic plate: An overview, in LeMasurier, W.E., and Thomson, J.W., eds., Volcanoes of the Antarctic plate and southern oceans: American Geophysical Union, Antarctic Research Series, v. 48, p. 1–19.

Leshin, L.A., 2000, Insights into Martian water reservoirs from analyses of Martian Meteorite QUE94201: Geophysical Research Letters, v. 27, no. 14, p. 2017–2020, doi: 10.1029/1999GL008455.

Lucchitta, B.K., 1993, Ice in the northern plains: Relic of a frozen ocean?: MSATT Workshop Martian Northern Plains (abs.): Lunar and Planetary Science Institute Technical Report, v. 93–04, Part 1, p. 9–10.

Lucchitta, B.K., Ferguson, H.M., and Summers, C., 1986, Sedimentary deposits in the northern lowland plains, Mars: Proceedings of the 17th Lunar and Planetary Science Conference, Part 1: Journal of Geophysical Research, v. 91, p. 166–174, doi: 10.1029/JB091iB13p0E166.

Lucchitta, B.K., Isbell, N.K., and Howington-Kraus, A., 1994, Topography of Valles Marineris: Implications for erosional and structural history: Journal of Geophysical Research, v. 99, p. 3783–3798, doi: 10.1029/93JE03095.

Major, J.J., and Newhall, C.G., 1989, Snow and ice perturbation during historical volcanic eruptions and the formation of lahars and floods: Bulletin of Volcanology, v. 52, p. 1–27, doi: 10.1007/BF00641384.

Malin, M.C., and Edgett, K.S., 2000, Evidence for Recent Groundwater Seepage and Surface Runoff on Mars: Science, v. 288, p. 2330–2335, doi: 10.1126/science.288.5475.2330.

Malin, M.C., Edgett, K.S., Posiolova, L.V., McColley, S.M., and Noe Dobrea, E.Z., 2006, Present-day impact cratering rate and contemporary gully activity on Mars: Science, v. 314, p. 1573–1577, doi: 10.1126/science.1135156.

Masursky, H., 1973, An overview of geologic results from Mariner 9: Journal of Geophysical Research, v. 78, p. 4037–4047.

Merrill, G.P., 1908, The Meteor Crater of Canyon Diablo, Arizona: Its history, origin and associated meteoric irons: Smithsonian Institution Miscellaneous Collections, v. L, p. 461–498.

Mouginis-Mark, P.J., Harris, A.J.L., and Rowland, S.K., 2007, Terrestrial analogs to the calderas of the Tharsis volcanoes on Mars, in Chapman, M.G., ed., The Geology of Mars: Evidence from Earth-based Analogs: Cambridge, UK, Cambridge University Press, p. 71–94.

NewScientist, 2005, Antarctic ice sheet is an "awakened giant": http://www.newscientist.com/article.ns?id=dn6962.

Nininger, H.H., 1956, Arizona's meteorite crater: Past–present–future: American Meteorite Museum Publication, Sedona, Arizona, 232 p.

Pardee, J.T., 1910, The Glacial Lake Missoula, Montana: The Journal of Geology, v. 18, p. 376–386.

Pardee, J.T., 1942, Unusual currents in Glacial Lake Missoula, Montana: Geological Society of America Bulletin, v. 53, p. 1569–1600.

Parker, T.J., Saunders, R.S., and Schneeberger, D.M., 1989, Transitional morphology in the west Deuteronilus Mensae region of Mars: Implications for modification of the lowland/upland boundary: Icarus, v. 82, p. 111–145, doi: 10.1016/0019-1035(89)90027-4.

Parker, T.J., Gorsline, D.S., Saunders, R.S., Pieri, D.C., and Schneeberger, D.M., 1993, Coastal geomorphology of the Martian northern plains: Journal of Geophysical Research, v. 98, p. 11,061–11,078, doi: 10.1029/93JE00618.

Pelletier, J.D., Kolb, K.J., McEwen, A.S., and Kirk, R.L., 2008, Recent bright gully deposits on Mars; wet or dry flow?: Geology, v. 36, no. 3, p. 211–214, doi: 10.1130/G24346A.1.

Petraglia, M., Korisettar, R., Boivin, N., Clarkson, C., Ditchfield, P., Jones, S., Koshy, J., Mirazón Lahr, M., Oppenheimer, C., Pyle, D., Roberts, R., Schwenninger, J.-L., Arnold, L., and White, K., 2007, Middle Paleolithic assemblages from the Indian Subcontinent before and after the Toba Supereruption: Science, v. 317, p. 114–116, doi: 10.1126/science.1141564.

Rampino, M.R., and Self, S., 1992, Volcanic winter and accelerated glaciation following the Toba Super-eruption: Nature, v. 359, p. 50–52, doi: 10.1038/359050a0.

Rampino, M.R., and Self, S., 1993, Climate-volcanism feedback and the Toba eruption of ~74,000 years ago: Quaternary Research, v. 40, p. 269–280, doi: 10.1006/qres.1993.1081.

Rampino, M.R., and Stothers, R.B., 1988, Flood basalt volcanism during the past 250 million years: Science, v. 241, p. 663–668, doi: 10.1126/science.241.4866.663.

Robinson, M.S., and Tanaka, K.L., 1990, Magnitude of a catastrophic flood event at Kasei Valles, Mars: Geology, v. 18, p. 902–905, doi: 10.1130/0091-7613(1990)018<0902:MOACFE>2.3.CO;2.

Rowley, R.J., Kostelnick, J.C., Braaten, D., and Li, X., 2007, Risk of rising sea level to population and land area: Eos, Transactions, American Geophysical Union, v. 88, no. 9, p. 105–107, doi: 10.1029/2007EO090001.

Sarna-Wojcicki, A.M., and Meyer, C.E., 1984, Tephrochronology applied to studies of Quaternary depositional basins in the western conterminous United States: Geological Society of America Abstracts with Programs, v. 16, no. 6, p. 644.

Scott, D.H., and Carr, M.H., 1978, Geologic map of Mars: United States Geological Survey Miscellaneous Investigation Series Map I-1083, 1:25,000,000 scale.

Scott, D.H., and Tanaka, K.L., 1986, Geologic map of the western equatorial region of Mars: United States Geological Survey Miscellaneous Investigation Series Map I-1802-A, 1:15,000,000 scale.

Scott, D.H., Chapman, M.G., Rice, J.W., Jr., and Dohm, J.M., 1992, New evidence of lacustrine basins on Mars: Amazonis and Utopia Planitiae: Proceedings of the 22nd Lunar and Planetary Science Conference, Lunar and Planetary Institute, Houston, v. 22, p. 53–62.

Self, S., Thordarson, T., and Widdowson, M., 2005, Gas fluxes from flood basalt eruptions: Elements, v. 1, no. 5, p. 283–287, doi: 10.2113/gselements.1.5.283.

Shoemaker, E.M., 1960, Impact mechanics at Meteor Crater Arizona [PhD Thesis]: New Jersey, Princeton University, unpublished, 55 p.

Smellie, J.L., 2000, Subglacial eruptions, in Sigurdsson, H., ed., Encyclopedia of Volcanoes: San Diego, Academic Press, p. 403–418.

Smellie, J.L., 2009, this volume, Terrestrial subice volcanism: Landform morphology, sequence characteristics, environmental influences, and implications for candidate Mars examples, in Chapman, M.G., and Keszthelyi, L.P., eds., Preservation of random megascale events on Mars and Earth: Influence on geologic history: Geological Society of America Special Paper 453, doi: 10.1130/2009.453(05).

Snorrason, Á., Jónsson, P., Sigurðsson, O., Pálsson, S., Árnason, S., Vikingsson, S., and Kaldal, I., 2002, November 1996 jökulhlaup on Skeiðarársandur outwash plain, Iceland, in Martini, I.P., Baker, V.R., and Garzón, G., eds., Flood and Megaflood Processes and Deposits: Recent and Ancient Examples: International Association of Sedimentologists Special Publication 32, p. 55–65.

Squyres, S.W., Grotzinger, J.P., Arvidson, R.E., Bell, J.F., III, Calvin, W.M., Christensen, P.R., Clark, B.C., Crisp, J.A., Farrand, W.H., Herkenhoff, K.E., Johnson, J.R., Klingelhoefer, G., Knoll, A.H., McLennan, S.M., McSween, H.Y., Morris, R.V., Rice, J.W., Rieder, R., and Soderblom, L.A., 2004, In situ evidence for an ancient aqueous environment at Meridiani Planum, Mars: Science, v. 306, no. 5702, p. 1709–1714, doi: 10.1126/science.1104559.

Tanaka, K.L., and Scott, D.H., 1987, Geologic Map of the Polar Regions of Mars: U.S. Geological Survey Miscellaneous Investigation Series Map I-1802-C, 1:15,000,000 scale.

Tanaka, K.L., Scott, D.H., and Greeley, R., 1992, Global Stratigraphy, in Kieffer, H.H. Jakosky, B.M. Snyder, C.W. and Matthews, M.S., eds, Mars: Tucson, University of Arizona Press, p. 345–382.

The Sunday Telegraph, 2006, http://www.telegraph.co.uk/news/main.jhtml?xml=/news/2006/12/10/nclimate10.xml&DCMP=EMC-new_10122006.

Thordarson, T., Rampino, M., Keszthelyi, L.P., and Self, S., 2009, this volume, Effects of megascale eruptions on Earth and Mars, in Chapman, M.G., and Keszthelyi, L.P., eds., Preservation of random megascale events on Mars and Earth: Influence on geologic history: Geological Society of America Special Paper 453, doi: 10.1130/2009.453(04).

Tilghman, B.C., 1905, Coon Butte, Arizona: Proceedings of the Academy of National Sciences, Philadelphia, v. LVII, p. 887–914.

Tweed, F.S., and Russell, A.J., 1999, Controls on the formation and sudden drainage of glacier-impounded lakes: Implications for jökulhlaup characteristics: Progress in Physical Geography, v. 23, p. 79–110.

Waitt, R.B., 2002, Great Holocence floods along Jökulsá á Fjöllum, north Iceland, in Martini, I.P., Baker, V.R., and Garzón, G., eds., Flood and Megaflood Processes and Deposits: Recent and Ancient Examples: International Association of Sedimentologists Special Publication, v. 32, p. 37–51.

Weis, P., and Neuman, W.L., 1971, The Channeled Scabland of eastern Washington: The geologic story of the Spokane flood: U.S. Geological Survey pamphlet.

Wignall, P.B., 2001, Large Igneous Provinces and mass extinctions: Earth-Science Reviews, v. 53, p. 1–33, doi: 10.1016/S0012-8252(00)00037-4.

Wilson, L., and Head, J.W., 1994, Review and analysis of volcanic eruption theory and relationships to observed landforms: Reviews of Geophysics, v. 32, p. 221–264, doi: 10.1029/94RG01113.

Witbeck, N.E., and Underwood, J.R., Jr., 1983, Geologic mapping in the Cydonia region of Mars: Reports of the Planetary Geologic Program: NASA Technical Memorandum 86246, p. 327–329.

Wood, C.A., and Kienle, J., eds., 1990, Volcanoes of North America: Cambridge, UK, Cambridge University Press, 354 p.

MANUSCRIPT ACCEPTED BY THE SOCIETY 3 NOVEMBER 2008

The Geological Society of America
Special Paper 453
2009

Effect of impact cratering on the geologic evolution of Mars and implications for Earth

Nadine G. Barlow*

Department of Physics and Astronomy, Northern Arizona University, Flagstaff, Arizona 86011-6010, USA

ABSTRACT

Impact cratering has affected the surfaces of all bodies in our Solar System. These short-duration but energetic events can drastically affect the regional and occasionally the global environment of a planet. The cratering record is better preserved on Mars than on Earth due to longer-term stability of the Martian crust and lower degradation rates. Impact cratering had its greatest effect early in Solar System history when bombardment rates were higher than today and the sizes of the impacting objects were larger. The record from this period of time is largely lost on Earth. High bombardment rates early in Solar System history may have eroded the Martian atmosphere to its present thin state, causing dramatic climate change. The regolith covering much of the Martian surface and the large quantities of dust seen in the atmosphere and covering much of the ground have been attributed to fragmentation of target material by impacts. Heating associated with crater formation may have contributed volatiles to the Martian atmosphere and initiated some of the outflow channels. The effects of an impact event extend far beyond the crater rim, and the planet's volatile-rich environment likely contributes to the greater ejecta extents seen on Mars than on the Moon. The cratering record of Mars thus holds important implications for how impacts may have affected the geologic evolution of Earth.

INTRODUCTION

Impact cratering is the only geologic process known to affect all bodies in our Solar System. Impacts into gaseous planets can produce short-lived atmospheric spots, as dramatically revealed by the 1994 Comet Shoemaker-Levy 9 impacts into Jupiter. However, it is the crater scars retained on surfaces of solid bodies which provide important insights on how impact processes have affected the geologic evolution of Solar System bodies.

Impact crater characteristics depend on the densities of the projectile and target material, impact velocity of the meteorite, and angle of impact (Melosh, 1989). The presence or absence of an atmosphere also affects the resulting crater morphology and provides insights into erosional regimes which have modified the crater. It is through detailed study of the morphologic and morphometric features associated with impact craters on different worlds that we can begin to understand the environmental conditions under which these craters formed.

Lunar craters are traditionally viewed as the standard by which to compare impact craters throughout the Solar System. This is because they were the first pristine craters to be studied in detail. However, when determining the effects that impact events have had on the terrestrial environment, Martian impact craters might be a better analog to use. Mars today has a thin atmosphere of carbon dioxide and no liquid water on its surface, but there is increasing evidence that its atmosphere may have been thicker

*nadine.barlow@nau.edu

and surface conditions may have been wetter (either as liquid water or ice) throughout much of the planet's past (Baker, 2001; Jakosky and Phillips, 2001). This is in contrast to the Moon, which shows no evidence of ever possessing an atmosphere or abundant surface/subsurface volatiles. Thus, effects produced by impacts into Earth's volatile-rich environment can be better understood through analysis of Martian impact craters.

THE LATE HEAVY BOMBARDMENT PERIOD ON MARS AND EARTH

Analysis of the number of craters within specific size ranges (called the *size-frequency distribution*) provides estimates of the formation ages of terrain units and information on how the size-frequency distribution of the impacting population may have changed over time. Comparison of crater size-frequency distributions among inner solar system bodies shows many similarities, suggesting that this region of the Solar System has been impacted by similar populations of impacting objects. As Figure 1 shows, the size-frequency distribution curves for the Earth, Venus, lunar maria, and Martian plains show a very flat distribution while those for the heavily cratered lunar highlands, Martian highlands, and observed surface of Mercury display a more complicated, multisloped shape.

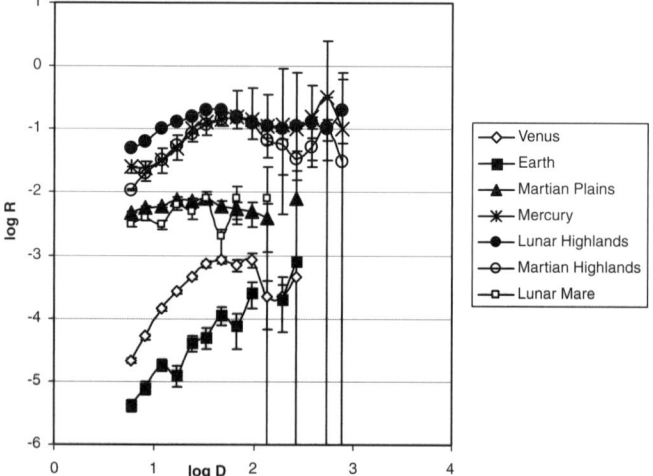

Figure 1. Crater size-frequency distribution data for inner solar system bodies, displayed on a Relative or R-Plot (see discussion in Barlow, 1990b). Heavily cratered regions of the Moon, Mercury, and Mars display high density multisloped curves while the lunar maria and Martian plains show a lower density, flatter curve. The steep decline in crater density at lower crater diameters on Earth and Venus is probably due to the active erosional environments of these bodies. Lunar highlands, lunar maria, and Mercury data are from Strom (1977). Martian data are from Barlow's *Catalog of Large Martian Impact Craters* (Barlow, 2006b). Venus data are from the Venus Crater Database by R. Herrick (www.lpi.usra.edu/resources/vc/vchome.html) and terrestrial data are from the Earth Impact Database (www.unb.ca/passc/ImpactDatabase/).

The typical explanation for the downturn in crater frequency on heavily cratered surfaces at diameters ≤70 km is obliteration of these craters by geologic processes. Lava flows, tectonism, fluvial and glacial erosion, eolian erosion, and ejecta deposition by subsequent impacts can destroy evidence of preexisting craters. Crater depth is proportional to crater diameter (Pike, 1980; Melosh, 1989; Garvin and Frawley, 1998; Garvin et al., 2000), so smaller craters are infilled and destroyed faster than larger craters. However, while some of the downturn in the crater size-frequency distribution curves is due to erosion, it is not the total story. Barlow (1990a) showed that craters still retaining distinct ejecta blankets (and thus not affected by large amounts of degradation) display this downturn on older Martian terrain units.

The downturn in the crater size-frequency distribution curves therefore results from two effects: erosion and differences in the size-frequency distributions of impacting populations responsible for crater formation (Chapman and McKinnon, 1986; Strom et al., 1992). The two different size-frequency distribution curves seen among inner solar system bodies indicate two different production populations (Barlow, 1990b; Strom et al., 1992). The heavily cratered surfaces of the Moon, Mercury, and Mars retain impact scars from leftover remnants of planetary accretion, a period called the late heavy bombardment period. Impact rates during this time period are estimated to have been up to 500 times higher than the present cratering rate (Barlow, 1990c). This late heavy bombardment period occurred after the planets had reached their current sizes and their crusts had solidified, as distinct from the early heavy bombardment period when the planets were still undergoing accretion. Large pieces of material still floated through space during this time, occasionally colliding with the newly formed planets to form large impact basins (craters typically > 200 km diam).

Lightly cratered surfaces seen on the lunar maria and the Martian plains display a crater size-frequency distribution which is comparable to that of present-day asteroids and comets. This is the post-heavy bombardment population of impacting objects which has dominated the inner solar system's cratering record for the past ~3.8×10^9 a, as indicated by analysis of lunar samples (Stöffler and Ryder, 2001). Earth and Venus show lower crater densities than the lunar maria and Martian plains, but craters on those bodies have been affected by large amounts of erosion.

Analysis of the cratering record indicates that large basin-forming impacts occurred primarily during late heavy bombardment. These large impacts are the ones which most dramatically affect the geologic evolution of a planet. For example, the discovery of buried basins in Mars Orbiter Laser Altimeter (MOLA) topography supports the theory that formation of large basins early in Martian history contributed to (although is likely not the sole explanation of [Zuber et al., 2000]) the formation of the ~3 km elevation difference between the southern highlands and northern plains (i.e., the "hemispheric dichotomy") (Frey and Schultz, 1988; Frey et al., 2002; Andrews-Hanna et al., 2008). Similarly, basin-forming impacts in the early history of Earth may have contributed to topographic differences between continents and ocean basins (Frey, 1977), particularly if there was a sudden increase in the number of basin-forming impacts near the end of the late heavy bombardment period as suggested by analysis of lunar samples (Stöffler and Ryder, 2001; Kring and Cohen, 2002) and dynamical considerations (Gomes et al., 2005; Strom et al., 2005). Early topographic variations induced by basin formation on Earth may have helped dictate the zones of weakness subsequently utilized by plate tectonic activity.

The Martian surface is covered by a layer of fragmented material called regolith. Although fluvial, glacial, and eolian erosional processes have contributed to regolith formation, impact cratering, particularly during the high impact rates of the late heavy bombardment period, was likely the primary source of this material (Hartmann et al., 2001). High impact rates on Earth also likely contributed to production of an early regolith, which has subsequently been modified by biologic and geologic activity and incorporated into the soils and sedimentary rocks which we see today. The Martian regolith is a major sink for volatiles from the atmosphere and previous episodes of a surface hydrosphere (Clark and Baird, 1979; Fanale and Jakosky, 1982; Donahue, 1995), and Earth's early regolith may have played a comparable role. Impact into such a regolith can release volatiles into the atmosphere, affecting the subsequent climatic and geologic evolution of the planet (Carr, 1989; Segura et al., 2002).

High impact rates and collisions with large meteoroids greatly affect a planet's atmosphere. Impact erosion, where passage of large meteoroids enhances the escape of a fraction of the atmosphere to space, is now generally accepted as a contributor to the loss of an early, denser atmosphere on Mars (Melosh and Vickery, 1989). Although impact erosion would be less of a factor for Earth's atmospheric density due to our planet's higher gravitational pull, very large impacts during late heavy bombardment could have resulted in the loss of some small fraction of our atmosphere to space (Vickery and Melosh, 1990; Newman et al., 1999). Even without any atmospheric loss, impacts can affect the atmosphere through the heating which causes chemical changes. Passage of the ~10-km-diam bolide that produced the Chixculub crater 65×10^6 a ago is believed to have produced large amounts of nitrogen oxides during its atmospheric passage, which could contribute to impact-induced acid rain formation (Lewis et al., 1982; Zahnle, 1990). Creation of such compounds not only affects any life on the planet's surface but also can alter weathering rates of surface materials.

Atmospheres also influence the distribution of material ejected during crater formation. Impacts onto airless bodies distribute ejecta along ballistic trajectories, modified by impact angle and the object's rotation. On bodies with an atmosphere, winds produced during impact events as well as disturbances to the planet's general atmospheric circulation patterns will distribute ejecta and any other products of the impact (e.g., soot from impact-induced wildfires) over large but often nonuniform areas. Detailed modeling of atmospheric influences on ejecta distribution have been conducted for both Earth (Shuvalov, 1999) and Mars (Nemtchinov et al., 2002; Cho and Stewart, 2003) and show major differences from the distribution of ejecta on airless bodies like the Moon.

There is little doubt that impacts affect the existence and evolution of life. There is now strong evidence that the mass extinction at the end of the Cretaceous Period 65×10^6 a ago was caused by the impact of a ~10-km-diam bolide into the Yucatan peninsula and surrounding continental shelf of present-day Mexico (Hildebrand et al., 1991; papers in Koeberl and MacLeod, 2002). Possible geologic evidence of impact-induced mass extinctions has also been reported at or near the Frasnian-Famennian (~367 Ma) (Claeys et al., 1992; Sandberg et al., 2002), Permian-Triassic (~260 Ma) (Becker et al., 2001; Erwin et al., 2002), Triassic-Jurassic (~205 Ma) (Bice et al., 1992; Olsen et al., 2002), Late Eocene–Early Oligocene (~36 Ma) (Alvarez et al., 1982; Ganapathy, 1982), and Pliocene (~2.3 Ma) (Kyte, 1988). Researchers also have noticed the coincidence between the onset of terrestrial life between 3.5×10^9 and 4.0×10^9 a ago with the end of late heavy bombardment (Barlow, 1990b; Chyba, 1993). Although life may have originated prior to the end of the heavy bombardment period (~3.8×10^9 a ago), the steady rain of meteorites and the formation of many large basins (particularly if the "lunar cataclysm" occurred) would have delayed widespread establishment of life until after the late heavy bombardment ended.

Comparison of inner solar system cratering records leads to some definite conclusions about how impacts have affected Earth's geologic evolution. Earth's cratering record is largely lost due to its active geologic environment, and the effect of impacts on Earth prior to 3×10^9 a must be gleaned from analysis of the cratering record on bodies where ancient terrain still survives. The numbers of terrestrial impact craters is low (~175) and most are eroded to the point where primary features are highly altered or completely destroyed. Erosion rates on Mars are much lower and impact craters there, which also formed in a volatile-rich environment, can provide important insights into the environmental consequences of terrestrial impacts.

MARTIAN IMPACT CRATERS: MORPHOLOGIES AND MORPHOMETRIES

Martian impact craters display a number of features that differ from their lunar counterparts and that likely result from the volatile-rich environment on Mars. Small (simple) impact craters on Mars display the traditional bowl-shaped appearance of simple craters elsewhere, while larger craters display the central peaks, wall terraces, shallower depths, and flatter floors associated with complex craters (Fig. 2). The diameter at which craters transition between simple and complex morphologies occurs at ~7 km on Mars, although this transition diameter varies with latitude, probably due to latitudinal variations in the concentration of subsurface ice (Garvin et al., 2000). The simple-to-complex transition diameter scales inversely with the planet's gravitational acceleration ($D_{sc} \propto 1/g$). Based on gravity considerations alone, the simple-to-complex transition diameter on Mars should be

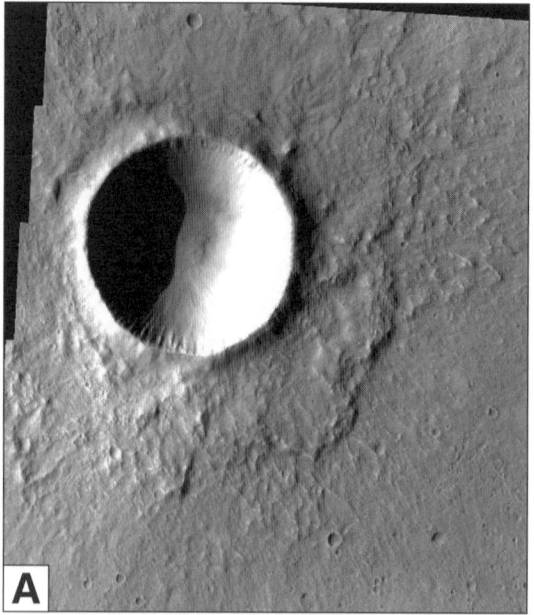

Figure 2. Simple and complex craters. (A) Example of a bowl-shaped simple crater on Mars. Crater is 6 km in diam and located at 36.2°N, 311.1°E (THEMIS image V04970003). (B) This 12-km-diam crater (13.3°N, 82.9°E) displays the central peak and wall terraces characteristic of complex craters (THEMIS image V01832003).

around 10 km, similar to that for smaller but denser Mercury. The lower observed value indicates that Martian surface materials are weaker than those seen on Mercury or the Moon, likely resulting from the presence of ice/water within the near-surface region.

One feature commonly seen in Martian impact craters but rarely seen on the Moon is central pits (Fig. 3). Pits may either occur as depressions in the center of the crater floor ("floor pits") or as small depressions at the top of a central rise or peak ("summit pits"). A Viking-based analysis by Barlow and Bradley (1990) identified ~1100 Martian impact craters containing central pits; a new analysis utilizing Mars Global Surveyor (MGS) Mars Orbiter Camera (MOC) and Mars Odyssey Thermal Emission Imagining System (THEMIS) imagery is revealing even larger numbers of central pit craters on Mars (Barlow and Hillman, 2006). Central pits are also seen on icy moons in the outer solar system (Schenk, 1993), leading to speculation that pits form by the sudden release of gases produced by vaporization of subsurface ice during crater formation. Pits are seen in Martian craters with a wide range of preservational states (which are related to crater age), indicating that subsurface volatiles have existed on the planet for much of its history (Barlow, 2006a). Recent two-dimensional (2D) and three-dimensional (3D) modeling of impacts into mixed ice-soil targets as well as pure ice targets shows that the highest temperatures are reached under the center of the transient crater floor, resulting in production of large amounts of vapor in this region (Pierazzo et al., 2005; Stewart and Senft, 2008). Release of this vapor could produce the central pits seen on Mars and the icy moons.

Most fresh Martian impact craters are surrounded by an ejecta blanket which has been emplaced by some type of fluidization process (Fig. 4A). These layered ejecta morphologies have been attributed to either impact into and vaporization of subsurface volatiles (Carr et al., 1977; Wohletz and Sheridan, 1983; Stewart et al., 2001) or interaction of an ejecta plume with the thin Martian atmosphere (Schultz and Gault, 1979; Schultz, 1992; Barnouin-Jha et al., 1999a, 1999b), although both mechanisms may play some role (Barlow, 2005a).

Martian impact craters display a wide variety of ejecta patterns (see discussion and references in Barlow et al., 2000). The three major types of layered ejecta morphologies are single layer (SLE; one complete ejecta layer surrounding the crater), double layer (DLE; two complete ejecta layers), and multiple layer (MLE; three or more partial or complete ejecta layers) ejecta morphologies (Fig. 4, A–C). Some Martian craters display only a lunarlike ("radial") ejecta pattern (Fig. 4D) with little or no indication of a fluidized component—these tend to be very small (<3-km-diam) or very large (>50-km-diam) craters. Other craters display secondary crater chains extending outward from the edge of a layered ejecta blanket (Mouginis-Mark et al., 2003; Block and Barlow, 2005) (Fig. 4E). Most of the layered ejecta morphologies terminate in a distal ridge ("rampart craters") (Fig. 4A), although others display a convex edge to their ejecta blanket ("pancake craters") (Fig. 4F). Some small craters on Mars are elevated above their surroundings due to later erosional processes ("pedestal craters") (Fig. 4G) (Arvidson et al., 1976; Head and

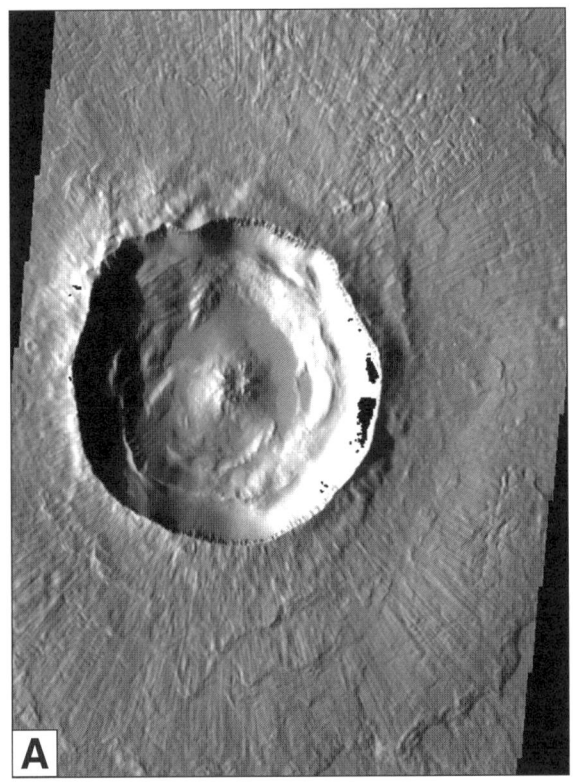
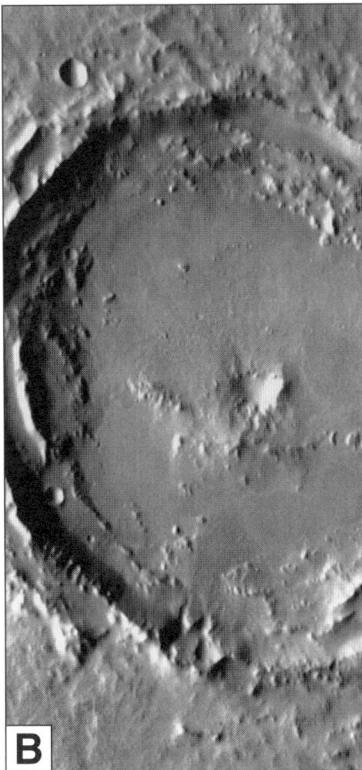

Figure 3. Martian central pit craters. (A) This 9.2-km-diam crater, located at 13.52°N, 287.35°E, displays a small summit pit atop its central peak (THEMIS image V05199007). (B) A central floor pit is visible in this 49.3-km-diam crater at 4.54°N, 320.76°E (THEMIS image I02374006).

Roth, 1976; Barlow, 2005b; Meresse et al., 2005; Kadish and Barlow, 2006; Kadish et al., 2008).

Several studies have shown that the smallest diameter at which a layered ejecta morphology is seen ("onset diameter") varies primarily as a function of latitude, although regional variations within particular latitude zones also are observed (Kuzmin et al., 1988; Demura and Kurita, 1998; Barlow et al., 2001; Reiss et al., 2006). The onset diameter for SLE craters is typically around 5 to 6 km near the equator, decreasing to <1 km near the poles. This latitudinal variation is consistent with the proposed distribution of near-surface ice based on geothermal and insolation models (Clifford, 1993) as well as near-surface ice maps produced by neutron analysis from Mars Odyssey's Gamma Ray Spectrometer (Boynton et al., 2002; Feldman et al., 2002, 2004; Mitrofanov et al., 2002, 2004). Such observations are among the strongest arguments for subsurface volatiles contributing to the formation of layered ejecta morphologies (Barlow, 2005a).

Ejecta blankets surrounding lunar impact craters are divided into "continuous" and "discontinuous" ejecta. The continuous ejecta blanket consists of thick, hummocky deposits typically within one crater radius of the rim. The discontinuous ejecta blanket is composed of thin, patchy deposits and includes secondary crater chains. The extent of discontinuous ejecta varies, but lunar crater rays often extend up to 10–20 crater radii. If layered ejecta deposits surrounding Martian impact craters correspond to lunar continuous ejecta, we find that Martian ejecta tend to travel further, probably due to lubrication of the flow provided by volatiles in the substrate and/or atmosphere. Ejecta flow extent for layered ejecta morphologies is typically expressed by the ejecta mobility (EM) ratio (Mouginis-Mark, 1979; Costard, 1989):

$$\text{EM} = \frac{\text{(maximum radial extent of ejecta)}}{\text{(crater radius)}}. \quad (1)$$

Analysis of 6000 layered ejecta craters (Barlow, 2001; Barlow and Pollak, 2002) reveals that the EM varies between 0.2 and 6.8 for SLE craters with an average value of 1.69. EM for MLE craters ranges between 0.3 and 5.2 with an average of 1.96. The inner layer of DLE craters has an average EM of 1.50 (range: 0.4–4.8), while the outer layer has an average EM of 3.01 (range 0.8–7.8). Thus ejecta deposits of Martian impact craters affect a larger region of surrounding terrain than those of lunar craters, in spite of the Moon's smaller gravity. EM of Martian impact craters also displays regional and latitudinal variations consistent with the proposed distribution of subsurface volatiles on Mars (e.g., EM becomes larger at higher latitudes).

Secondary crater chains extending beyond the layered ejecta deposits are the Martian equivalent of lunar discontinuous ejecta deposits. Secondary cratering on Mars appears to be heterogeneous, depending on crater size and terrain unit on which the crater is superposed. A survey of craters displaying secondary crater chains reveals that craters as small as 10 km diam show secondary

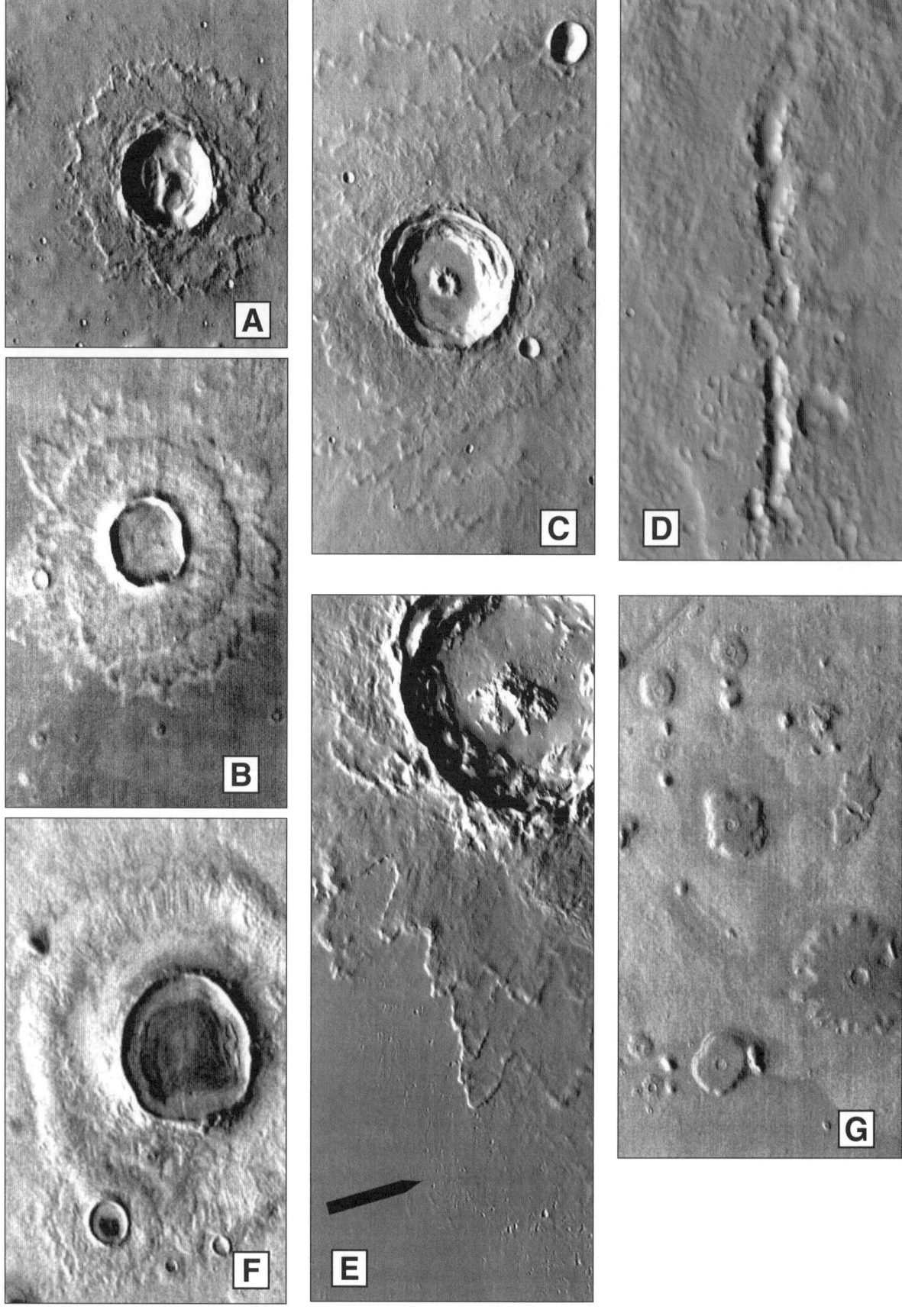

Figure 4. Examples of Martian impact crater ejecta morphologies. (A) This 11.6-km-diam crater (23.63°N, 101.71°E) is an example of a single layer ejecta (SLE) crater. The ejecta blanket terminates in a distal ridge, leading this to be classified as a single layer ejecta rampart (SLER) crater (THEMIS image I11555003). (B) Double layer ejecta (DLE) craters, such as this 10.0-km-diam example at 53.27°N, 283.14°E, show two distinct ejecta layers (THEMIS image I11786003). (C) Multiple layer ejecta (MLE) craters display three or more partial or complete ejecta layers. This 15.5-km-diam example at 11.46°N, 138.70°E also displays a prominent central pit (THEMIS image I10418013). (D) Secondary crater chains are a distinguishing feature of radial ejecta (Rd) craters. These secondaries are south of a 113.3-km-diam crater centered at 28.77°N, 355.22°E (THEMIS image I09849023). (E) MOC and THEMIS images have revealed craters displaying secondary crater chains (arrowed) extending just beyond the edge of a layered ejecta morphology. This example is associated with a 28.3-km-diam crater at 23.19°N, 207.76°E (THEMIS image I01990002). (F) Pancake craters do not show the distal ridge characteristic of rampart craters. Instead, the edge of the ejecta blanket slopes downward from the top of the flow layer. The inner layer of DLE craters displays many of the same morphological characteristics as pedestal craters and the two crater types may be related. This pancake crater is 10.2 km in diam and is centered at 38.38°N, 177.91°E (THEMIS image I03177002). (G) Pedestal craters have both the ejecta and crater perched above the surroundings. These unique crater forms likely result from postimpact erosional processes. This cluster of pedestal craters is located near 42.5°N, 154.4°E. The largest crater in this group (lower right side of image) is 2.8 km in diam (THEMIS image I11416010).

crater chains on young (Amazonian-aged) volcanic plains while craters have to be >45 km diam on old (Noachian-aged) highlands units (Hartmann and Barlow, 2006). According to computer modeling, secondary craters are produced from high-velocity material derived from a narrow boundary zone, called the spallation zone, during transient crater formation (Melosh, 1989; Head et al., 2002; Artemieva and Ivanov, 2004). Hartmann and Barlow (2006) propose that strength of the surface material dictates whether secondary crater chains are produced. Impact into competent lava flows produces spalls of sufficient size and energy to produce secondary craters even with moderately sized impacts. In the highlands, eons of impact cratering have produced the fragmented regolith (Hartmann et al., 2001). Only large impacts excavate through this regolith and encounter the underlying competent layer with enough energy to produce coherent material necessary for secondary crater formation.

Secondary craters on Mars also travel considerable distances from the primary crater. THEMIS infrared imagery clearly shows thermally distinct rays extending from a few recent primary craters (McEwen et al., 2005; Tornabene et al., 2005). Secondary craters along rays from the 10-km-diam Zunil crater have been detected up to 1600 km (320 crater radii) from the primary crater (McEwen et al., 2005), considerably further than ray material on the Moon. These observations suggest that the majority of craters <1 km diam may be secondary craters. This result not only affects crater-based age estimates for the Martian surface but also has important implications for the extent of the region affected by an individual impact.

Secondary cratering also may be responsible for some Martian crater clusters which do not at first appearance seem to be associated with any primary impact crater. Popova et al. (2003) have identified two types of "isolated" crater clusters: small clusters, with craters ranging in diameter to a few tens of meters and spread over an area of a few hundred meters, and large clusters whose craters display diameters up to a few hundred meters and which are spread over regions up to a few tens of km in extent. Modeling by Popova et al. (2003) suggest that small crater clusters result from fragmentation of weak meteoroids during passage in the Martian atmosphere. Large crater clusters, however, appear to be secondary craters produced by material ejected along suborbital trajectories by primary impacts a few tens of km in diameter (Popova et al., 2007).

IMPLICATIONS FOR EARTH

The Earth's active erosional environment eliminates pristine features associated with impact craters on a geologically short time scale. Although recent craters such as Meteor Crater (USA), Ries (Germany), and Chicxulub (Mexico) still retain some surrounding ejecta deposits, their eroded nature makes it difficult to determine whether they were emplaced by ballistic or flow processes (Hörz et al., 1983; Newsom et al., 1986; Grant and Schultz, 1993; Pope et al., 2005). However, the presence of an atmosphere and volatile-rich surface materials suggest that layered ejecta deposits, like those seen on Mars, probably formed around at least some terrestrial impact craters (Kenkmann and Schönian, 2006). Detailed studies of craters formed in ice and/or water-rich environments (e.g., Grieve, 1988; Masaitis, 1999; Tsikalas et al., 1998; Horton et al., 2005) may reveal remnants of features similar to those seen in Martian layered ejecta deposits. Based on the Mars analysis, we can expect that both the continuous and discontinuous ejecta deposits associated with terrestrial impacts affected a larger area around the crater than expected from lunar extrapolations. There is some indication of this with Chicxulub ejecta, which has been detected at distances over 2 crater radii away from the impact site (Pope et al., 2005).

Researchers also should be alert to evidence for central pits in terrestrial impact craters, particularly those which formed in known volatile-rich environments. Only three terrestrial craters to date show evidence of a central uplift topped by a pit which could be a summit pit (Puchezh-Katunk in Russia [Masaitis, 1999; Barlow et al., 2007], Mulkarra in Australia [Plescia, 1999], and Obolon in Ukraine [Schmieder and Buchner, 2008]), but others may exist and are not being recognized due to the lunar crater bias. Central depressions, depressions atop central peaks, and peak rings seen in craters smaller than expected should be investigated more thoroughly for the possibility that these are terrestrial examples of central pit craters.

SUMMARY

Analysis of impact craters and the cratering record on Mars provides important insights into how impact cratering has likely affected Earth's geologic evolution. The high impact rates and large basin-forming impacts that occurred during late heavy bombardment, prior to ~3.8×10^9 a ago, appear to have contributed to the Martian hemispheric dichotomy, regolith formation, and atmospheric evolution. Earth likely experienced similar evolutionary effects caused by impacts during the late heavy bombardment period. Martian impact craters display a range of morphologies that differ from lunar impact craters because of the presence of an atmosphere and near-surface volatiles. Layered ejecta blankets, central pits, and more extensive ejecta extent are apparently the result of impact into the volatile-rich environment of Mars. Terrestrial impact craters, particularly those which formed in water and/or ice-rich environments, should be carefully investigated for possible evidence of similar features.

ACKNOWLEDGMENTS

The Mars crater analysis discussed in this paper was supported under NASA Mars Data Analysis Awards NAG5-8265 and NAG5-12510. The author thanks Mary Chapman for a careful review of the paper and the MOC and THEMIS investigations for the spectacular new crater information they are providing.

REFERENCES CITED

Alvarez, W., Alvarez, L.W., Asaro, F., and Michel, H.V., 1982, Iridium anomaly approximately synchronous with terminal Eocene extinctions: Science, v. 216, p. 886–888, doi: 10.1126/science.216.4548.886.

Andrews-Hanna, J.C., Zuber, M.T., and Banerdt, W.B., 2008, The Borealis basin and the origin of the Martian crustal dichotomy: Nature, v. 453, p. 1212–1215.

Artemieva, N.A., and Ivanov, B., 2004, Launch of Martian meteorites in oblique impacts: Icarus, v. 171, p. 84–101, doi: 10.1016/j.icarus.2004.05.003.

Arvidson, R.E., Coradini, M., Carusi, A., Coradini, A., Fulchignoni, M., Federico, C., Funiciello, R., and Salomone, M., 1976, Latitudinal variation of wind erosion of crater ejecta deposits on Mars: Icarus, v. 27, p. 503–516, doi: 10.1016/0019-1035(76)90166-4.

Baker, V.R., 2001, Water and the Martian landscape: Nature, v. 412, p. 228–236, doi: 10.1038/35084172.

Barlow, N.G., 1990a, Constraints on early events in Martian history as derived from the cratering record: Journal of Geophysical Research, v. 95, p. 14191–14201, doi: 10.1029/JB095iB09p14191.

Barlow, N.G., 1990b, Application of the inner Solar System cratering record to the Earth, in Sharpton, V. L. and Ward, P. D., eds., Global Catastrophes in Earth History: An Interdisciplinary Conference on Impacts, Volcanism, and Mass Mortality: Boulder, Colorado, Geological Society of America Special Paper 247, p. 181–187.

Barlow, N.G., 1990c, The late heavy bombardment crater size-frequency distribution function in the inner solar system: International Workshop on Meteorite Impact on the Early Earth: Houston, Texas, Lunar and Planetary Institute, LPI Contribution Number 746, p. 4–5.

Barlow, N.G., 2001, Ejecta mobility results for impact craters in the northern hemisphere of Mars: Lunar and Planetary Science XXXII: Houston, Texas, Lunar and Planetary Institute, Abstract no. 1606.

Barlow, N.G., 2005a, A review of Martian impact crater ejecta structures and their implications for target properties, in Kenkmann, T., Hörz, F., and Deutsch, A., eds., Large Meteorite Impacts III: Boulder, Colorado, Geological Society of America Special Paper 384, p. 433–442.

Barlow, N.G., 2005b, A new model for pedestal crater formation: Workshop on the Role of Volatiles and Atmospheres on Martian Impact Craters: Houston, Texas, Lunar and Planetary Institute, LPI Contribution no. 1273, p. 15–16.

Barlow, N.G., 2006a, Impact craters in the northern hemisphere of Mars: Layered ejecta and central pit characteristics: Meteoritics and Planetary Science, v. 41, p. 1425–1436.

Barlow, N.G., 2006b, Status report on the "Catalog of Large Martian Impact Craters," version 2.0: Lunar and Planetary Science XXXVII: Houston, Texas, Lunar and Planetary Institute, Abstract no. 1337.

Barlow, N.G., and Bradley, T.L., 1990, Martian impact craters: Correlations of ejecta and interior morphologies with diameter, latitude, and terrain: Icarus, v. 87, p. 156–179, doi: 10.1016/0019-1035(90)90026-6.

Barlow, N.G., and Pollak, A., 2002, Comparisons of ejecta mobility ratios in the northern and southern hemispheres of Mars: Lunar and Planetary Science XXXIII: Houston, Texas, Lunar and Planetary Institute, Abstract no. 1322.

Barlow, N.G., Koroshetz, J., and Dohm, J.M., 2001, Variations in the onset diameter for Martian layered ejecta morphologies and their implications for subsurface volatile reservoirs: Geophysical Research Letters, v. 28, p. 3095–3098, doi: 10.1029/2000GL012804.

Barlow, N.G., Sharpton, V.L., and Kuzmin, R.O., 2007, Impact structures on Earth and Mars, in Chapman, M., ed., The Geology of Mars: Evidence from Earth-Based Analogs: Cambridge, UK, Cambridge University Press, p. 47–70.

Barlow, N.G., Boyce, J.M., Costard, F.M., Craddock, R.A., Garvin, J.B., Sakimoto, S.E.H., Kuzmin, R.O., Roddy, D.J., and Soderblom, L.A., 2000, Standardizing the nomenclature of Martian impact crater ejecta morphologies: Journal of Geophysical Research, v. 105, p. 26733–26738, doi: 10.1029/2000JE001258.

Barnouin-Jha, O.S., Schultz, P.H., and Lever, J.H., 1999a, Investigating the interactions between an atmosphere and an ejecta curtain. 1. Wind tunnel tests: Journal of Geophysical Research, v. 104, p. 27105–27115, doi: 10.1029/1999JE001026.

Barnouin-Jha, O.S., Schultz, P.H., and Lever, J.H., 1999b, Investigating the interactions between an atmosphere and an ejecta curtain. 2. Numerical experiments: Journal of Geophysical Research, v. 104, p. 27117–27131, doi: 10.1029/1999JE001027.

Becker, L., Poreda, R.J., Hunt, A.G., Bunch, T.E., and Rampino, M., 2001, Impact event at the Permian-Triassic Boundary: Evidence from extraterrestrial noble gases in fullerenes: Science, v. 291, p. 1530–1534, doi: 10.1126/science.1057243.

Bice, D.M., Newton, C.R., McCauley, S., Reiners, P.W., and McRoberts, C.A., 1992, Shocked quartz at the Triassic-Jurassic boundary in Italy: Science, v. 255, p. 443–446, doi: 10.1126/science.255.5043.443.

Block, K.M., and Barlow, N.G., 2005, Secondary cratering rates on the basaltic plains of the Moon and Mars: Lunar and Planetary Science XXXVI: Houston, Texas, Lunar and Planetary Institute, Abstract no. 1816.

Boynton, W.V., Feldman, W.C., Squyres, S.W., Prettyman, T.H., Brücker, J., Evans, L.G., Reedy, R.C., Starr, R., Arnold, J.R., Drake, D.M., Englert, P.A.J., Metzger, A.E., Mitrofanov, I., Trombka, J.I., d'Uston, C., Wänke, H., Gasnault, O., Hamara, D.K., Janes, D.M., Marcialis, R.L., Maurice, S., Mikheeva, I., Taylor, G.J., Tokar, R., and Shinohara, C., 2002, Distribution of hydrogen in the near-surface of Mars: Evidence for subsurface ice deposits: Science, v. 297, p. 81–85, doi: 10.1126/science.1073722.

Carr, M.H., 1989, Recharge of the early atmosphere of Mars by impact-induced release of CO_2: Icarus, v. 79, p. 311–327, doi: 10.1016/0019-1035(89)90080-8.

Carr, M.H., Crumpler, L.S., Cutts, J.A., Greeley, R., Guest, J.E., and Masursky, H., 1977, Martian impact craters and emplacement of ejecta by surface flow: Journal of Geophysical Research, v. 82, p. 4055–4065, doi: 10.1029/JS082i028p04055.

Chapman, C.R., and McKinnon, W.K., 1986, Cratering of planetary satellites, in Burns, J. A. and Matthews, M. S., eds., Satellites: Tucson, Univ. Arizona Press, p. 492–580.

Cho, J.Y.-K., and Stewart, S.T., 2003, Global dispersal of dust following impact cratering events on Mars: Lunar and Planetary Science XXXIV: Houston, Texas, Lunar and Planetary Institute, Abstract no. 2101.

Chyba, C.F., 1993, The violent environment of the origin of life: Progress and uncertainties: Geochemical et Cosmochimicia Acta, v. 57, p. 3351–3358, doi: 10.1016/0016-7037(93)90543-6.

Claeys, P., Casier, J.-G., and Margolis, S.V., 1992, Microtektites and mass extinctions: Evidence for a Late Devonian asteroid impact: Science, v. 257, p. 1102–1104, doi: 10.1126/science.257.5073.1102.

Clark, B.C., and Baird, A.K., 1979, Volatiles in the Martian regolith: Geophysical Research Letters, v. 6, p. 811–814, doi: 10.1029/GL006i010p00811.

Clifford, S.M., 1993, A model for the hydrologic and climatic behavior of water on Mars: Journal of Geophysical Research, v. 98, p. 10973–11016, doi: 10.1029/93JE00225.

Costard, F.M., 1989, The spatial distribution of volatiles in the Martian hydrolithosphere: Earth, Moon, and Planets, v. 45, p. 265–290, doi: 10.1007/BF00057747.

Demura, H., and Kurita, K., 1998, A shallow volatile layer at Chryse Planitia, Mars: Earth, Moon, and Planets, v. 50, p. 423–429.

Donahue, T.M., 1995, Evolution of water reservoirs on Mars from D/H ratios in the atmosphere and crust: Nature, v. 374, p. 432–434, doi: 10.1038/374432a0.

Erwin, D.H., Bowring, S.A., and Yugan, J., 2002, End-Permian mass extinctions: A review, in Koeberl, C., and MacLeod, K.G., eds., Catastrophic Events and Mass Extinctions: Impacts and Beyond: Boulder, Colorado, Geological Society of America Special Paper 356, p. 363–383.

Fanale, F.P., and Jakosky, B.M., 1982, Regolith-atmosphere exchange of water and carbon dioxide on Mars: Effects on atmospheric history and climate change: Planetary and Space Science, v. 30, p. 819–831, doi: 10.1016/0032-0633(82)90114-3.

Feldman, W.C., Boynton, W.V., Tokar, R.L., Prettyman, T.H., Gasnault, O., Squyres, S.W., Elphic, R.C., Lawrence, D.J., Lawson, S.L., Maurice, S., McKinney, G.W., Moore, K.R., and Reedy, R.C., 2002, Global distribution of neutrons from Mars: Results from Mars Odyssey: Science, v. 297, p. 75–78, doi: 10.1126/science.1073541.

Feldman, W.C., Prettyman, T.H., Maurice, S., Plaut, J.J., Bish, D.L., Vaniman, D.T., Mellon, M.T., Metzger, A.E., Squyres, S.W., Karunatillake, S., Boynton, W.V., Elphic, R.C., Funsten, H.O., Lawrence, D.J., and Tokar, R.L., 2004, Global distribution of near-surface hydrogen on Mars: Journal of Geophysical Research, v. 109, doi: 10.1029/2003je002160.

Frey, H., 1977, Origin of the Earth's ocean basins: Icarus, v. 32, p. 235–250, doi: 10.1016/0019-1035(77)90064-1.

Frey, H.V., and Schultz, R.A., 1988, Large impact basins and the mega-impact origin for the crustal dichotomy on Mars: Geophysical Research Letters, v. 15, p. 229–232, doi: 10.1029/GL015i003p00229.

Frey, H.V., Roark, J.H., Shockey, K.M., Frey, E.L., and Sakimoto, S.E.H., 2002, Ancient lowlands on Mars: Geophysical Research Letters, v. 29, doi: 10.1029/2001GL013832.

Ganapathy, R., 1982, Evidence for a major meteorite impact on the Earth 34 million years ago: Implications for Eocene extinctions: Science, v. 216, p. 885–886, doi: 10.1126/science.216.4548.885.

Garvin, J.B., and Frawley, J.J., 1998, Geometric properties of Martian impact craters: Preliminary results from the Mars Orbiter Laser Altimeter: Geophysical Research Letters, v. 25, p. 4405–4408, doi: 10.1029/1998GL900177.

Garvin, J.B., Sakimoto, S.E.H., Frawley, J., and Schnetzler, C., 2000, North polar region craterforms on Mars: Geometric characteristics from the Mars Orbiter Laser Altimeter: Icarus, v. 144, p. 329–352, doi: 10.1006/icar.1999.6298.

Gomes, R., Levison, H.F., Tsiganis, K., and Morbidelli, A., 2005, Origin of the cataclysmic late heavy bombardment period of the terrestrial planets: Nature, v. 435, p. 466–469, doi: 10.1038/nature03676.

Grant, J.A., and Schultz, P.H., 1993, Erosion of ejecta at Meteor Crater, Arizona: Journal of Geophysical Research, v. 98, p. 15033–15047, doi: 10.1029/93JE01580.

Grieve, R.A.F., 1988, The Haughton impact structure: Summary and synthesis of the results of the HISS project: Meteoritics, v. 23, p. 249–254.

Hartmann, W.K., and Barlow, N.G., 2006, Nature of the Martian uplands and the Martian meteorite age distribution: Meteoritics & Planetary Science, v. 41, p. 1453–1467.

Hartmann, W.K., Anguita, J., de la Casa, M.A., Berman, D.C., and Ryan, E.V., 2001, Martian cratering 7: The role of impact gardening: Icarus, v. 149, p. 37–53, doi: 10.1006/icar.2000.6532.

Head, J.N., Melosh, H.J., and Ivanov, B.A., 2002, Martian meteorite launch: High-speed ejecta from small craters: Science, v. 298, p. 1752–1756, doi: 10.1126/science.1077483.

Head, J.W., and Roth, R., 1976, Mars pedestal crater escarpments: Evidence for ejecta related emplacement: Papers Presented to the Symposium on Planetary Cratering Mechanics, p. 50–52.

Hildebrand, A.R., Penfield, G.T., Kring, D.A., Pilkington, M., Camargo, Z.A., Jacobsen, S.B., and Boynton, W.V., 1991, Chicxulub Crater: A possible Cretaceous-Tertiary boundary impact crater on the Yucatan Peninsula, Mexico: Geology, v. 19, p. 867–871, doi: 10.1130/0091-7613(1991)019<0867:CCAPCT>2.3.CO;2.

Horton, J.W., Aleinikoff, J.N., Kunk, M.J., Gohn, G.S., Edwards, L.E., Self-Trail, J.M., Powars, D.S., and Izett, G.A., 2005, Recent research on the Chesapeake Bay impact structure, USA—Impact debris and reworked ejecta, in Kenkmann, T., Hörz, F., and Deutsch, A., eds., Large Meteorite Impacts III: Boulder, Colorado, Geological Society of America Special Paper 384, p. 147–170.

Hörz, F., Ostertag, R., and Rainey, D.A., 1983, Bunte Breccia of the Ries: Continuous deposits of large impact craters: Reviews of Geophysics and Space Physics, v. 21, p. 1667–1725, doi: 10.1029/RG021i008p01667.

Jakosky, B.M., and Phillips, R.J., 2001, Mars volatile and climate history: Nature, v. 412, p. 237–244, doi: 10.1038/35084184.

Kadish, S.J., and Barlow, N.G., 2006, Pedestal crater distribution and implications for a new model of formation: Lunar and Planetary Science XXXVII: Houston, Texas, Lunar and Planetary Institute, Abstract no. 1254.

Kadish, S.J., Head, J.W., Barlow, N.G., and Marchant, D.R., 2008, Martian pedestal craters: Marginal sublimation pits implicate a climate-related formation mechanism: Geophysical Research Letters, v. 35, CiteID L16104, doi: 10.1029/2008GL034990.

Kenkmann, T., and Schönian, F., 2006, Ries and Chicxulub: Impact craters on Earth provide insights for Martian ejecta blankets: Meteoritics and Planetary Science, v. 41, p. 1587–1603.

Koeberl, C., and MacLeod, K.G., 2002, Catastrophic Events and Mass Extinctions: Impacts and Beyond: Boulder, Colorado, Geological Society of America Special Paper 356, 746 pp.

Kring, D.A., and Cohen, B.A., 2002, Cataclysmic bombardment throughout the inner solar system 3.9 – 4.0 Ga: Journal of Geophysical Research, v. 107, doi: 10.1029/2001JE001529.

Kuzmin, R.O., Bobina, N.N., Zabalueva, E.V., and Shashkina, V.P., 1988, Structural inhomogeneities of the Martian cryosphere: Solar System Research, v. 22, p. 121–133.

Kyte, F.T., 1988, The extraterrestrial component in marine sediments: Description and interpretation: Paleoceanography, v. 3, p. 235–247, doi: 10.1029/PA003i002p00235.

Lewis, J.S., Watkins, G.H., Hartman, H., and Prinn, R.G., 1982, Chemical consequences of major impact events on Earth, in Silver, L.T., and Schultz, P.H., eds., Geological implications of impacts of large asteroids and comets on the Earth: Boulder, Colorado, Geological Society of America Special Paper 190, p. 215–221.

Masaitis, V.L., 1999, Impact structures of northeastern Eurasia: The territories of Russia and adjacent countries: Meteoritics & Planetary Science, v. 34, p. 691–711.

McEwen, A.S., Preblich, B.S., Turtle, E.P., Artemieva, N.A., Golombek, M.P., Hurst, M., Kirk, R.L., Burr, D.M., and Christensen, P.R., 2005, The rayed crater Zunil and interpretations of small impact craters on Mars: Icarus, v. 176, p. 351–381, doi: 10.1016/j.icarus.2005.02.009.

Melosh, H.J., 1989, Impact Cratering: A Geologic Process: New York, Oxford University Press, 245 pp.

Melosh, H.J., and Vickery, A.M., 1989, Impact erosion of the primordial atmosphere of Mars: Nature, v. 338, p. 487–489, doi: 10.1038/338487a0.

Meresse, S., Baratoux, D., Costard, F., and Mangold, N., 2005, Perched craters and episodes of sublimation on northern plains: Workshop on the Role of Volatiles and Atmospheres on Martian Impact Craters: Houston, Texas, Lunar and Planetary Institute, LPI Contribution no. 1273, p. 75–76.

Mitrofanov, I., Anfimov, D., Kozyrev, A., Litvak, M., Sanin, A., Tret'yakov, V., Krylov, A., Shvetsov, V., Boynton, W.V., Shinohara, C., Hamara, D., and Saunders, R.S., 2002, Maps of subsurface hydrogen from the High Energy Neutron Detector, Mars Odyssey: Science, v. 297, p. 78–81.

Mitrofanov, I.G., Litvak, M.L., Kozyrev, A.S., Sanin, A.B., Tret'yakov, V.I., Grin'kov, V.Yu., Boynton, W.V., Shinohara, C., Hamara, D., and Saunders, R.S., 2004, Soil water content on Mars as estimated from neutron measurements by the HEND instrument onboard the 2001 Mars Odyssey spacecraft: Solar System Research, v. 38, p. 253–257.

Mouginis-Mark, P., 1979, Martian fluidized ejecta morphology: Variations with crater size, latitude, altitude, and target material: Journal of Geophysical Research, v. 84, p. 8011–8022, doi: 10.1029/JB084iB14p08011.

Mouginis-Mark, P.J., Boyce, J.M., Hamilton, V.E., and Anderson, F.S., 2003, A very young, large, impact crater on Mars: Sixth International Conference on Mars: Houston, Texas, Lunar and Planetary Institute, Abstract no. 3004.

Nemtchinov, I.V., Shuvalov, V.V., and Greeley, R., 2002, Impact-mobilized dust in the martian atmosphere: Journal of Geophysical Research, v. 107, doi: 10.1029/2001JE001834.

Newman, W.I., Symbalisty, E.M.D., Ahrens, T.J., and Jones, E.M., 1999, Impact erosion of planetary atmospheres: Some surprising results: Icarus, v. 138, p. 224–240, doi: 10.1006/icar.1999.6076.

Newsom, H.E., Graup, G., Sewards, T., and Keil, K., 1986, Fluidization and hydrothermal alteration of the suevite deposit at the Reis Crater, West Germany, and implications for Mars: Proceedings of the 17th Lunar and Planetary Science Conference, Part 1: Journal of Geophysical Research, v. 91, p. E239–E251, doi: 10.1029/JB091iB13p0E239.

Olsen, P.E., Koeberl, C., Huber, H., Montanari, A., Fowell, S.J., Et-Touhami, M., and Kent, D.V., 2002, Continental Triassic-Jurassic boundary in central Pangea: Recent progress and discussion of an Ir anomaly, in Koeberl, C., and MacLeod, K. G., eds., Catastrophic Events and Mass Extinctions: Impacts and Beyond: Boulder, Colorado, Geological Society of America Special Paper 356, p. 505–522.

Pierazzo, E., Artemieva, N.A., and Ivanov, B.A., 2005, Starting conditions for hydrothermal systems underneath Martian craters: Hydrocode modeling, in Kenkmann, T., Hörz, F., and Deutsch, A., eds., Large Meteorite Impacts III: Boulder, Colorado, Geological Society of America Special Paper 384, p. 443–457.

Pike, R.J., 1980, Control of crater morphology by gravity and target type: Mars, Earth, Moon, in Proceedings of the 11th Lunar and Planetary Science Conference: New York, Pergamon Press, p. 2159–2189.

Plescia, J.B., 1999, Mulkarra impact structure, South Australia: A complex impact structure: Lunar and Planetary Science XXX: Houston, Texas, Lunar and Planetary Institute, Abs. no. 1889.

Pope, K.O., Ocampo, A.C., Fischer, A.G., Vega, F.J., Ames, D.E., King, D.T., Fouke, B.W., Wachtman, R.J., and Kletetschka, G., 2005, Chicxulub impact ejecta deposits in southern Quintana Roo, Mexico, and central Belize, in Kenkmann, T. Hörz, and Deutsch, A., eds., Large Meteorite Impacts III: Boulder, Colorado, Geological Society of America Special Paper 384, p. 171–190.

Popova, O.P., Nemtchinov, I.V., and Hartmann, W.K., 2003, Bolides in the present and past Martian atmosphere and effects on cratering processes: Meteoritics & Planetary Science, v. 38, p. 905–925.

Popova, O.P., Hartmann, W.K., Nemtchinov, I.V., Richardson, D.C., and Berman, D.C., 2007, Crater clusters on Mars: Shedding light on martian ejecta launch conditions: Icarus, v. 190, p. 50–73.

Reiss, D., van Gasselt, S., Hauber, E., Michael, G., Jaumann, R., and Neukum, G., 2006, Ages of rampart craters in equatorial regions of Mars: Implications for the past and present distribution of ground ice: Meteoritics and Planetary Science, v. 41, p. 1437–1452.

Sandberg, C.A., Morrow, J.R., and Ziegler, W., 2002, Late Devonian sea-level changes, catastrophic events, and mass extinctions, in Koerberl, C., and MacLeod, K. G., eds., Catastrophic Events and Mass Extinctions: Impacts and Beyond: Boulder, Colorado, Geological Society of America Special Paper 356, p. 473–487.

Schenk, P.M., 1993, Central pit and dome craters: Exposing the interiors of Ganymede and Callisto: Journal of Geophysical Research, v. 98, p. 7475–7498, doi: 10.1029/93JE00176.

Schmieder, M., and Buchner, E., 2008, Dating impact craters: Palaeogeographic versus isotopic and stratigraphic methods—a brief case study: Geological Magazine, v. 145, p. 586–590.

Schultz, P.H., 1992, Atmospheric effects on ejecta emplacement: Journal of Geophysical Research, v. 97, p. 11623–11662.

Schultz, P.H., and Gault, D.E., 1979, Atmospheric effects on Martian ejecta emplacement: Journal of Geophysical Research, v. 84, p. 7669–7687.

Segura, T.L., Toon, O.B., Colaprete, A., and Zahnle, K., 2002, Environmental effects of large impacts on Mars: Science, v. 298, p. 1977–1980, doi: 10.1126/science.1073586.

Shuvalov, V.V., 1999, Atmospheric plumes created by meteoroids impacting the Earth: Journal of Geophysical Research, v. 104, p. 5877–5890, doi: 10.1029/1998JE900037.

Stewart, S.T., and Senft, L.E., 2008, Advances in modeling collisions on icy bodies: Large meteorite Impacts and Planetary Evolution IV: Houston, Texas, Lunar Planetary Institute, Abs. 3085.

Stewart, S.T., O'Keefe, J.D., and Ahrens, T.J., 2001, The relationship between rampart crater morphologies and the amount of subsurface ice: Lunar and Planetary Science XXXII: Houston, Texas, Lunar and Planetary Institute, Abstract no. 2092.

Stöffler, D., and Ryder, G., 2001, Stratigraphy and isotope ages of lunar geologic units: Chronological standard for the inner Solar System, in Kallenbach, R., et al., eds., Chronology and Evolution of Mars: Dordrecht, Kluwer Academic Publishers, p. 9–54.

Strom, R.G., 1977, Origin and relative age of lunar and Mercurian intercrater plains: Physics of the Earth and Planetary Interiors, v. 15, p. 156–172, doi: 10.1016/0031-9201(77)90028-0.

Strom, R.G., Croft, S.K., and Barlow, N.G., 1992, The Martian impact cratering record, in Kieffer, H.H., et al., eds., Mars: Tucson, Univ. Arizona Press, p. 383–423.

Strom, R.G., Malhotra, R., Takashi, I., Yoshida, F., and Kring, D.A., 2005, The origin of planetary impactors in the inner solar system: Science, v. 309, p. 1847–1850, doi: 10.1126/science.1113544.

Tornabene, L.L., Mcweeen, H.Y., Moersch, J.E., Piatek, J.L., Milam, K.A., and Christensen, P.R., 2005, Recognition of rayed craters on Mars in THEMIS thermal infrared imagery: Implications for Martian meteorite source regions: Lunar and Planetary Science XXXVI: Houston, Texas, Lunar and Planetary Institute, Abstract no. 1970.

Tsikalas, F., Gudlaugsson, S.T., and Faleide, J.I., 1998, The anatomy of a buried complex impact structure: The Mjølnir structure, Barents Sea: Journal of Geophysical Research, v. 103, p. 30469–30483, doi: 10.1029/97JB03389.

Vickery, A.M., and Melosh, H.J., 1990, Atmospheric erosion and impactor retention in large impacts, with applications to mass extinctions, in Sharpton, V.L., and Ward, P.D., eds., Global catastrophes in Earth history: An interdisciplinary conference on impacts, volcanism, and mass mortality: Boulder, Colorado, Geological Society of America Special Paper 247, p. 289–300.

Wohletz, K.H., and Sheridan, M.F., 1983, Martian rampart crater ejecta: Experiments and analysis of melt-water interaction: Icarus, v. 56, p. 15–37, doi: 10.1016/0019-1035(83)90125-2.

Zahnle, K.J., 1990, Atmospheric chemistry by large impacts, in Sharpton, V.L., and Ward, P.D., eds., Global catastrophes in Earth history: An interdisciplinary conference on impacts, volcanism, and mass mortality: Boulder, Colorado, Geological Society of America Special Paper 247, p. 271–288.

Zuber, M.T., Solomon, S.C., Phillips, R.J., Smith, D.E., Tyler, G.L., Aharonson, O., Balmino, G., Banerdt, W.B., Head, J.W., Johnson, C.L., Lemoine, F.G., McGovern, P.J., Neumann, G.A., Rowlands, D.D., and Zhong, S., 2000, Internal structure and early thermal evolution of Mars from Mars Global Surveyor topography and gravity: Science, v. 287, p. 1788–1793, doi: 10.1126/science.287.5459.1788.

MANUSCRIPT ACCEPTED BY THE SOCIETY 3 NOVEMBER 2008

Megafloods and global paleoenvironmental change on Mars and Earth

Victor R. Baker*

Department of Hydrology and Water Resources, University of Arizona, Tucson, Arizona 85721-0011, USA

ABSTRACT

The surface of Mars preserves landforms associated with the largest known water floods. While most of these megafloods occurred more than 1 Ga ago, recent spacecraft images document a phase of outburst flooding and associated volcanism that seems no older than tens of millions of years. The megafloods that formed the Martian outflow channels had maximum discharges comparable to those of Earth's ocean currents and its thermohaline circulation. On both Earth and Mars, abrupt and episodic operations of these megascale processes have been major factors in global climatic change. On relatively short time scales, by their influence on oceanic circulation, Earth's Pleistocene megafloods probably (1) induced the Younger Dryas cooling of 12.8 ka ago, and (2) initiated the Bond cycles of ocean-climate oscillation with their associated Heinrich events of "iceberg armadas" into the North Atlantic. The Martian megafloods are hypothesized to have induced the episodic formation of a northern plains "ocean," which, with contemporaneous volcanism, led to relatively brief periods of enhanced hydrological cycling on the land surface (the "MEGAOUTFLO Hypothesis"). This process of episodic short-duration climate change on Mars, operating at intervals of hundreds of millions of years, has parallels in the Neoproterozoic glaciation of Earth (the "Snowball Earth Hypothesis"). Both phenomena are theorized to involve abrupt and spectacular planet-wide climate oscillations, and associated feedbacks with ocean circulation, land-surface weathering, glaciation, and atmospheric carbon dioxide. The critical factors for megascale environmental change on both Mars and Earth seem to be associated tectonics and volcanism, plus the abundance of water for planetary cycling. Some of the most important events in planetary history, including those of the biosphere, seem to be tied to cataclysmic episodes of massive hydrological change.

INTRODUCTION

Earth and Mars are the only two planets known to have geological histories involving vigorous surface cycling of water between reservoirs of ice, liquid, and vapor. Although the current state of the Martian surface is exceedingly cold and dry, there are extensive reservoirs of polar ice and ground ice (Boynton et al., 2002), plus probable immense quantities of groundwater beneath an ice-rich permafrost zone (Clifford, 1993; Clifford and Parker, 2001). Results from recent spacecraft missions have generally corroborated a theory, first presented by Baker et al. (1991), that periodically Mars's cold, dry, stable state has been perturbed by massive outbursts of water and gas, leading to a cool, wet atmosphere in contact with standing bodies of water (lakes, seas, or even a temporary "ocean"). Labeled "the episodic ocean hypothesis" by Carr (1996), the Baker et al. (1991) theory was initially considered to be outrageous (Kerr, 1993). The prevailing theoretical view of Mars until relatively recently has been that Mars Is

*baker@hwr.arizona.edu

continuously Dead and Dry Except during the Noachian (MIDDEN). (The Noachian is the "early Mars" epoch, extending from 4.5 to ~3.8 Ga ago). The MIDDEN hypothesis is finally being recognized as inconsistent with numerous important aspects of Martian geology, including extensive evidence for very recent water-related landforms (e.g., Baker, 2001, 2004, 2005). In light of the new evidence, the original Baker et al. (1991) hypothesis has been regenerated as MEGAOUTFLO: "Mars Episodic Glacial Atmospheric Oceanic Upwelling by Thermotectonic FLood Outburst" (Baker et al., 2000; see also, Kargel, 2004).

On Earth, the oceans are integral to long-term climate change. Their currents distribute heat between the equator and the poles. For example, the Gulf Stream flows at discharges of up to 100 sverdrups (Sv), involving relatively slow-moving water 1 km in depth and 50–75 km wide. (The sverdrup, equivalent to 1×10^6 m^3/s, is the unit of discharge for both ocean currents and megafloods.) As northward-flowing Atlantic Ocean seawater evaporates and becomes more saline, it increases in density, eventually sinking in the northern Atlantic. This process forms a portion of the great global thermohaline circulation pattern, which acts as a conveyor belt for heat in the oceans. However, this global-scale circulation pattern was disrupted during the last glaciation when megafloods (floods with peak discharges on the order of a Sv or more) introduced relatively low-density, freshwater lids over large areas of ocean surface. The resulting disruption of sea-surface temperatures and density structure drastically altered the meridional transport of heat on a global scale. Earth's global climate was altered on time scales of decades to centuries (Broecker et al., 1989; Barber et al., 1999). Thus both Earth and Mars have aspects of their long-term hydrological cycles that involve the cataclysmic-flood triggering of major epochs of global paleoenvironmental change.

MEGAFLOODS

High-energy megafloods are planetary-scale phenomena, associated with glacier outburst settings, glacial lake spillways, and the immense outflow channels of Mars (Baker, 2002). Paleohydraulic analyses of these floods show that their peak discharges can be comparable to those of ocean currents (Table 1). Collectively, discharges from all the Martian outflow channels may have totaled as much as 100–1000 Sv (Baker et al., 1991). These values imply total released reservoir volumes of 10^5–10^7 km^3 of water, using scaling relationships for terrestrial superfloods (Baker et al., 1993). The higher volumes from these preliminary calculations match the volume of water estimated from Martian topography to have ponded on the surface of the planet, mainly in a temporary "ocean" on its northern plains (Parker et al., 1993; Head et al., 1999).

Martian Outburst Flooding

The Martian outflow channels involve the immense upwelling of cataclysmic flood flows from subsurface sources (Baker and Milton, 1974), mostly during later periods of Martian history, after termination of the heavy bombardment phase at ~3.9 Ga. The huge peak discharges implied by the size and morphology of the outflow channels (Baker et al., 1992) are explained by several models (Table 2). A popular view is that a warm, wet climate phase during the heavy bombardment was followed by a progressively thickening ice-rich permafrost zone during subsequent cold and dry Martian history. The outflow channels result from releases of subsurface water that was confined by this process (Carr 1979, 2000; Clifford and Parker 2001). Certainly, there is strong evidence, notably from impact crater morphologies (Carr 1996), that much of the Martian surface is underlain by a thick ice-rich permafrost zone, a "cryolithosphere" (Kuzmin et al. 1988). However, the geological record shows that volcano-ice-water interactions are commonly associated with outburst flood channels. Thus episodic heat flow and volcanism (Baker et al. 1991) affords an alternative to the linear model of progressive pressurization of confined water by cryosphere thickening. The MEGAOUTFLO hypothesis envisions long periods (perhaps on the order of 10^8 a) during which Mars has a stable atmosphere that is cold and dry like that of today, with nearly all its water trapped as ground ice and underlying groundwater. The stable state is punctuated by relatively short-duration (perhaps 10^4 or 10^5 a) episodes of quasistable conditions that are warmer and wetter than those at present.

The energy for the immense outflow channel floods could have been supplied by the thermal and tectonic effects of immense mantle plumes. In this MEGAOUTFLO scenario, gas hydrate in the Martian permafrost zone is destabilized by episodes of very high heat flow (and associated volcanism), thereby releasing radiatively active gas from the lower permafrost zone (2 to 3 km depths). This process could involve carbon dioxide or methane (Kargel et al., 2000; Max and Clifford, 2001; Tanaka et al., 2001). The dissolved gas from the underlying groundwater and the gas released from ice in the permafrost zone would contribute to explosively pressurized slurries of water and sediments in massive outbursts (Komatsu et al., 2000). In the atmosphere these radiatively active gases can generate a transient greenhouse that would coincide with the immense pondings of water produced by the outburst floods.

Exceptionally young outflow channels and associated volcanism occur in both the Cerberus plains and the Tharsis regions of Mars (Hartmann and Berman, 2000; Mouginis-Mark 1990; Tanaka et al., 2005). Data from the Mars Global Surveyor (MGS) show that localized water releases, interspersed with lava flows, occurred approximately within the last 10 Ma (Berman and Hartmann 2002, Burr et al. 2002). The huge discharges associated with these floods and the temporally related volcanism should have introduced considerable water into active hydrological circulation on Mars. The remarkable evidence for very recent outburst floods (Burr et al., 2002) suggests that a relatively small MEGAOUTFLO episode occurred recently on Mars. Evidence for extensive sulfate salts on the Martian surface (Paige, 2005) is also consistent with this model, in that the SO_2 and water generated from

TABLE 1. PROPERTIES OF VARIOUS MEGAFLOODS AND MEGAFLOWS

Location	Width (km)	Depth (m)	Velocity (m/s)	Discharge (Sv)	References
Missoula Flood (Rathdrum)	6	150	22	20	Baker (1973); O'Connor and Baker (1992)
Altai Flood (Central Asia)	2.5	400	25	20	Baker et al. (1993)
Maadim Vallis (Mars)	5	100	10	5	Irwin et al. (2004)
Mangala Vallis (Mars)	14	100	15	20	Komar (1979)
Kasei Vallis (Mars)	83	374	32	1000	Robinson and Tanaka (1990)
Gulf Stream	50–75	1000	1–3	100	Gross (1987)

TABLE 2. MEGAFLOOD GENERATION (MARS)

Mechanism	Examples	References
Subice volcanism	Chryse Channels	Masursky et al. (1977); Chapman et al. (2003)
Lake spillways	Maadim Vallis	Irwin et al. (2004)
	Mangala Vallis	Zimbelman et al. (1992)
Hydrostatic pressure in large-scale confined aquifers	Chryse Channels	Carr (1979, 2000)
Liquefaction and mudflows	Ravi Vallis	Nummedal and Prior (1981)
Pressurized water release	Mangala Vallis	Tanaka and Chapman (1990)
From fractures, graben, dikes	Athabasca Vallis	Burr et al. (2002)
Decompression of CO_2	General	Hoffman (2000)
Decompression/melting of CO_2 and expulsion of water	General	Milton (1974); Komatsu et al. (2000); Max and Clifford (2001)

TABLE 3. MEGAFLOOD GENERATION (EARTH)

Mechanism	Example	Discharge	References
Subglacial volcanism	Icelandic Jokulhlaups	0.1–1.0	Bjornsson (2002)
Lake spillways	Bonneville	1	O'Connor (1993)
	Strait of Dover	10	Smith (1985)
Ice-dammed lake failure	Missoula	10–20	O'Connor and Baker (1992)
	Altai	10–20	Baker et al. (1993); Herget (2005)
	Agassiz	0.3	Teller et al. (2002)
	Agassiz, Ojibway	5–10	Clarke et al. (2004)
Pressurized subglacial outburst	Livingstone Event (Laurentide Ice Sheet)	60	Shaw (1996)

the volcanism and flood outburst would have produced the acid conditions to generate extensive sulfate salt emplacement on the Martian surface. The recent MEGAOUTFLO episode may also have been a trigger for water migration, leading to ice emplacement (Baker, 2003) in latitudinal zonation for which orbital variations likely acted as the pacemaker (Head et al., 2003).

Terrestrial Glacial Floods

Some Earth megafloods also derive from ice-volcano interactions (Table 3). However, ice-marginal lakes are the sources of the largest, well-documented (and noncontroversial) glacial megafloods. Perhaps the most famous example is Glacial Lake Missoula, which formed south of the Cordilleran Ice Sheet, in the northwestern United States (Baker, 1973). The Purcell Lobe of the Ice Sheet extended south from British Columbia to the basin of modern Pend Oreille Lake in northern Idaho. It thereby impounded the Clark Fork River drainage to the east, forming Glacial Lake Missoula in western Montana. At maximum extent this ice-dammed lake achieved a depth of 600 m at its dam, which held back a volume of ~2500 km^3 for which the water surface covered an area of 7500 km^2. Cataclysmic failure of the ice dam impounding this lake resulted in discharges up to ~20 Sv (O'Connor and Baker, 1992). On land these flows produced the distinctive erosional and depositional features of the Channeled Scabland, including scabland bedrock erosion, streamlining of residual uplands, large-scale scour around obstacles, depositional bars, giant current ripples, and huge sediment fans (Baker, 1981).

Upon reaching the Pacific Ocean, the Missoula floodwaters continued flowing down the continental slope as hyperpycnally generated turbidity currents. The sediment-charged floodwaters followed the Cascadia submarine channel into and through the Blanco Fracture Zone and out onto the abyssal plain of the Pacific. As much as 5000 km^3 of sediment may have been carried and distributed as turbidites over a distance of 2000 km west of the Columbia River mouth (Normark and Reid, 2003).

The largest well-documented glacial megalake formed in north-central North America in association with the largest glacier of the last ice age, the Laurentide Ice Sheet. This ice sheet is known to have been highly unstable throughout much of its history. Not only did freshwater discharges from the glacier result in ice-marginal lakes; outbursts of meltwater into the Atlantic Ocean may have generated climate changes by influencing the thermohaline circulation of the North Atlantic Ocean (Teller et al., 2002). As the Laurentide Ice Sheet of central and eastern Canada retreated from its late Quaternary maximum extent, it was bounded to the south and west by immense meltwater lakes, which developed in the troughs that surrounded the ice. As the lake levels rose, water was released as great megafloods, which carved numerous spillways into the drainages of the Mississippi, St. Lawrence, and Mackenzie Rivers (Kehew and Teller, 1994). The greatest megafloods developed from the last of the ice-marginal lakes, a union between Lake Agassiz in south-central Canada and Lake Ojibway in northern Ontario. The resulting megalake held ~160,000 km^3 of water, which was released subglacially ~8200 a ago under the ice sheet and into Labrador Sea via the Hudson Strait (Clarke et al., 2004).

A Laurentide meltwater-discharge event around 13,350 yr B.P. used the Hudson River valley. This megaflood drained Glacial Lakes Iroquois, Vermont, and Albany by breaching a moraine dam at the Narrows in New York City. It also initiated a short-lived (<400 a) cold event by diminishing the thermohaline circulation (Donnelly et al., 2005). The common influence of terrestrial megafloods is to generate more glacial conditions than what was prevailing immediately prior to the flooding.

Lowland glacial lakes of Siberia involved the expansion of huge Quaternary ice sheets that covered the shallow seas north of Eurasia and blocked north-flowing rivers, notably the Ob, Irtysh, and the Yenesei. In the 1970s Mikhail G. Grosswald recognized that, as in North America, the resulting proglacial lakes were immense and involved important diversions of meltwater through huge spillways (Grosswald, 1980). In European Russia, water that previously flowed to the Arctic was instead conveyed to the south-flowing Dnieper and Volga Rivers. More importantly, the great north-flowing Siberian Rivers, the Irtysh, Ob, and Yenesei, emptied into a megalake, Lake Mansi, which drained southward through the Turgai divide of north-central Kazakhstan to the basin of the modern Aral Sea. The Aral, in turn, drained though a spillway at its southwestern end into the Caspian, which expanded to a Late Quaternary size over twice its modern extent, known as the Khvalyn paleolake. The Khvalyn paleolake spilled westward through the Manych spillway into the Don valley, then to the Sea of Azov, and through the Kerch Strait to the Euxine Abyssal Plain (the floor of the modern Black Sea). Then, disconnected from the Agean and filled by freshwater (the New Euxine phase), the glacial meltwater filled the Black Sea basin. This basin functioned in two modes during the Quaternary. Its cold-climate mode was a freshwater lake (that may have filled and drained through the northwestern Turkey to the Agean). Its warm-climate phase involved rising global sea level, inducing salt-water invasions through the Turkish straits to form the saline Black Sea, the last of which occurred ~8000 a ago, an inundation that may have inspired the story of Noah (Ryan et al., 2003).

Subglacial Megaflood Hypotheses

Over the past 20 a, a controversial theory has emerged for immense cataclysmic flooding beneath the continental ice sheets of the last terrestrial glaciation. As outlined by Shaw (1996) this theory derives from inferred origins for a variety of enigmatic subglacial landforms, involving water erosion and deposition. These landforms include drumlins (Shaw, 1983, 2002; Shaw et al., 1989), Rogen moraines (Fisher and Shaw, 1993), large-scale bedrock erosional flutings and streamlining (Kor et al., 1991), gravel sheets in eskers (Brennand and Shaw, 1994), hummocky terrain (Munro and Shaw, 1997), pendant bars (Sjogren and Rains, 1995), and tunnel channels (valleys) (Shaw, 1988). Though commonly explained by subglacial ice deformational processes and related, small-scale water flows, the genesis of these features cannot be observed in modern glaciers that are much smaller than their late Quaternary counterparts. Shaw (1996) explains the assemblage of landforms as part of an erosional/depositional sequence beneath continental ice sheets that precedes regional ice stagnation and esker formation with a phase of immense subglacial sheet floods, which, in turn, follows ice-sheet advances that terminate with surging, stagnation, and melt-out. Shaw (1996) proposes that peak discharges of tens of sverdrups are implied by the late Quaternary subglacial landscapes of the southern Laurentide Ice Sheet. Release volumes are also huge. Blanchon and Shaw (1995) propose that a 14 m sea-level rise at 15 ka resulted from such megaflooding.

Shoemaker (1995) provides some theoretical support for Shaw's model, though at smaller flow magnitude levels. Arguments against the genetic hypotheses for the landforms are given by Benn and Evans (1998). Walder (1994) criticizes the hydraulics, and Clarke et al. (2005) object that the theory requires unreasonably huge volumes of meltwater to be subglacially or supraglacially stored and then suddenly released.

Subglacial cataclysmic flooding has recently been proposed for the Mid-Miocene ice sheet that overrode the Transantarctic Mountains (Denton and Sugden, 2005). Spectacular scabland erosion occurs on plateau and mountain areas up to 2100 m in elevation, over an 80-km mountain front. The inferred high-energy flooding could have been supplied from subglacial lakes, such as modern Lake Vostok (Siegert, 2005). Such lakes would have formed beneath the thickest, warm-based portion of the ice sheet, located inland (west) of the mountains (Sugden and Denton, 2004). Thinner ice overlying the mountains was probably cold-based, thereby generating a seal that could be broken when the pressure of water in the lakes, confined by thick overlying ice, reached threshold values. The water from the lakes would move along the ice-rock interface, along the pressure gradient up and across the mountain rim. The high-velocity flooding would be further enhanced to catastrophic proportions as conduits were opened by frictional heat in the subglacial flows (Denton and Sugden, 2005).

Perhaps the most spectacular hypothesized subglacial flooding is Grosswald's (1999) proposal that much of central Russia was inundated in the late Quaternary by immense outbursts from the ice-sheet margins to the north. Using satellite imagery to map large-scale streamlined topography and flowlike lineations, Grosswald (1999) infers that the entire Arctic Ocean was confined beneath an ice sheet that grew from converging ice shelves, which coalesced from the ice sheets that formed on the northern margins of the surrounding continents. This water then broke out during the Pleistocene from a point in northwestern Siberia, flowing toward the Caspian and then to the Black Sea. The pathway is approximately similar to that of the Turgai and Manych Spillways, noted above. However, this hypothesis, which has some similarities to proposals for subglacial megaflooding, has not been adequately tested in the field.

GLOBAL CONSEQUENCES

Earth's Oceans

Earth's oceans constitute a vast, interconnected body of water that covers ~70% of the planet's surface. Moreover, this water body has persisted throughout at least the last 3.9 Ga of planetary history, acting as a stabilizing force in the planet's climate engine. Nevertheless, important changes occur during Earth's ice ages, when large volumes of water transfer from the ocean to solid ice accumulation on the land.

The connections between Pleistocene ice sheets and the oceans are still very poorly understood. An important link is inferred from relatively brief (100- and 500-a) intervals in which thick marine layers of ice-rafted material were widely distributed across the North Atlantic. Called "Heinrich Events," these layers are thought to record episodes of massive iceberg discharge from unstable ice sheets (Bond and Lotti, 1995). The youngest Heinrich events date to 17,000 (H1), 24,000 (H2), and 31,000 (H3) a ago. Closely related is the Younger Dryas (YD) event, a global cooling at 12,800 a ago, which is also associated with North Atlantic ice-rafted rock fragments. There is evidence from corals at Barbados that sea level rose spectacularly, ~20 m during H1 and ~15 m just after YD. Such rises would require short-term freshwater fluxes to the oceans, respectively, of ~14,000 and 9000 km^3/a (Fairbanks, 1989). These events are thought to relate to ice-sheet collapse, reorganization of ocean-atmosphere circulation, and release of subglacial and proglacial meltwater, most likely during episodes of cataclysmic megaflooding.

Mars's "Oceans"

Evidence for persistent standing bodies of water on Mars is abundant (Chapman, 1994; Scott et al., 1995; Cabrol and Grin, 1999, 2001). Despite the lack of direct geomorphological evidence that the majority of Mars's surface was ever covered by standing water, the term "ocean" has been applied to temporary ancient inundations of the northern plains, which did not persist through the whole history of the planet. Although initially inferred from sedimentary landforms on the northern plains (e.g., Lucchitta et al., 1986), inundation of the northern plains has been most controversially tied to identifications of "shorelines" made by Parker et al. (1989, 1993). Failure to confirm some of the shoreline landforms on the newer Mars Orbiter Camera (MOC) imagery led some to reject the ocean hypothesis (Malin and Edgett, 1999, 2001), though it is only various shoreline interpretations that can be rejected in this manner, not the hypothesis of plains inundation. Nevertheless, the MGS data confirm the initial observations of a regionally mantling layer of sediment, now called the Vastitas Borealis Formation, covering perhaps 3 × 10^7 km^2 of the northern plains (Head et al., 2002). This sediment is contemporaneous with the post-Noachian outflow channels, and it was likely emplaced as the sediment-laden outflow channel discharges became hyperpycnal flows upon entering ponded water on the plains (Ivanov and Head, 2001). In another scenario, Clifford and Parker (2001) envision a Noachian "ocean," contemporaneous with the highlands valley networks and fed by a great fluvial system extending from the south polar cap, through Argyre and the Chryse Trough, to the northern plains.

The MEGAOUTFLO model ascribes the episodic formation of Oceanus Borealis on the northern plains of Mars as merely one component of a cycle. Phenomenally long epochs of post-heavy-bombardment time during which the Mars surface had extremely cold and dry conditions are punctuated by short-duration (~10^4–10^5 a) episodes of quasistable conditions, considerably wetter and somewhat warmer than those prevailing today at the planet's surface. The transition from the long-persistent cold-dry state was induced by cataclysmic outbursts of huge flood flows from the Martian outflow channels.

Megaflood-Ocean Interactions

Terrestrial megafloods influence the global climate system through their interactions with the ocean. A megaflood can enter the ocean either (1) as a buoyant spreading freshwater plume over higher density (salty) seawater (Kourafalou et al., 1996), or (2) as a descending flow of sediment-laden, high-density fluid, known as a hyperpycnal flow (Mulder et al., 2003). The turbidity current deposits of hyperpycnal flows may extend across hundreds, even thousands of kilometers of abyssal plain seafloor, as in the case of the Missoula Floods of the Cordilleran Ice Sheet (Normark and Reid, 2003). Nevertheless, these flows gradually lose momentum as they drop their sediment loads, thereby releasing low-density freshwater from the bottom of deep ocean basins. The buoyant freshwater will then move upward in massive convective plumes (Hesse et al., 2004). These will disrupt the thermal structure of the ocean, with consequences for the currents that distribute heat and moderate climates.

Sediment-charged Martian floods from outflow channels (Baker, 1982) would have entered the ponded water body on the northern plains as powerful turbidity currents. This is the reason for the lack of obvious deltalike depositional areas at the mouths of the outflow channels. High-velocity floods, combined with the effect of the reduced Martian gravity (lowering the settling velocities for entrained sediment) promotes unusually coarse-grained washload (Komar, 1980), permitting the turbidity currents to sweep over the entire northern plains. The latter are mantled by a vast deposit, the Vastitas Borealis Formation, which covers almost 3×10^7 km^2, or approximately one-sixth of the planet's area. This sediment is contemporaneous with the post-Noachian outflow channels, and it was likely emplaced as the sediment-laden outflow channel discharges became hyperpycnal flows upon entering ponded water on the plains (Ivanov and Head, 2001). In another scenario, Clifford and Parker (2001) envision a Noachian "ocean," contemporaneous with the highlands valley networks, and fed by a great fluvial system extending from the south polar cap, through Argyre and the Chryse Trough, to the northern plains.

GLOBAL PALEOENVIRONMENTAL CHANGE

Today, Mars is too cold and dry to have an Earth-like climate as a stable condition. Its stable state is not to have liquid water on its surface. Earth may also have experienced Mars-like climate phases in its remote past, but these were not stable climatic states. The stable condition for Earth is to have its ocean and atmosphere in continual contact, thereby moderating the climate. Even during its stable periods, such as the current glacial/interglacial cycling, Earth has been subject to short-term climatic changes of considerable magnitude. As noted above, massive floods of freshwater into the oceans are definitely a factor in cataclysmic climate change for the Quaternary Earth. However, this influence also interacts with many other processes that are only partially understood. Among these are (1) releases of methane from gas hydrates, stored in the sediments of the upper continental slopes of the seafloor (Kennett et al., 2000); and (2) changes in atmospheric carbon dioxide (Pierrehumbert, 2004); changes in ice-sheet dimensions (Stuut et al., 2004); and changes in biological productivity (Dugdale et al., 2004). It is clear that feedback processes are involved in the mutual interactions of these and other components of the global climate system. Could such feedback processes ever have driven Earth to ultraextreme climates?

"Icehouse" and "Greenhouse" Earth

In the 1960s atmospheric scientists began employing relatively simple energy-balance climate models to investigate feedback mechanisms for Earth's climate system. One important problem they encountered was the "faint young sun paradox," wherein solar irradiance is known to be increasing through time as a consequence of the thermonuclear evolution of the Sun. If the early Earth received much less irradiance, perhaps ~30% less during its first billion years (Gough, 1981), its surface should have become totally frozen. The initial growth of ice sheets induced by lower solar irradiance will be subsequently enhanced by the ice-albedo effect on climate change (Budyko, 1966). This positive feedback creates a fundamental instability (Erikson, 1968; Budyko, 1969; Sellers, 1969) such that a "runaway" cooling occurs after about half Earth's surface becomes ice covered.

The 1960s climate modelers knew geological evidence indicated that, at least generally for the last 3.9 Ga, Earth has had liquid oceans and related aqueous processes on its surface. The paradox is further enhanced by the presumption that Earth's frozen state would be irreversible. The ice-covered Earth would have such a high albedo that even the progressive increase in solar irradiance would never be able to bring mean surface temperatures above freezing. In essence this frozen Earth would have a climate very similar to that of present-day Mars. Without the moderating influence of ocean-atmosphere interactions, annual temperatures everywhere would remain below freezing. As on Mars, however, there would be huge diurnal and seasonal temperature oscillations, so that local summer temperatures would reach afternoon highs above freezing. The cycling of water would be extremely limited, dominated by sublimation instead of evaporation, even though immense reservoirs of water would be present beneath the frozen surfaces of the oceans.

The resolution to the "faint young sun paradox" is generally believed to be the greenhouse effect of carbon dioxide. Over long geological time scales Earth's "greenhouse" is presumably stabilized by the interaction of plate tectonics with the global carbon cycle (Walker et al., 1981). Carbon dioxide released through volcanoes continually adds to the atmosphere, but it is removed by precipitation over the continents. The dissolved carbon dioxide produces a slightly acidic precipitation, which reacts with silicate minerals, thereby generating mobile metal cations, silica, and bicarbonate. These eventually get precipitated, commonly in the shells and tests of marine organisms. Subduction of the seafloor, enriched in silica and carbonates, generates metamorphic reactions in Earth's mantle, liberating carbon dioxide, which rises with the magmas that subsequently degas at volcanoes, thereby completing the cycle. This long-term carbon cycle seems to operate with sufficient efficiency for Earth to remain in a "greenhouse" state, avoiding the "icehouse" state predicted by the "faint young sun paradox."

"Snowball Earth" Events

Although Earth never entered a permanent "icehouse" state, it did experience major epochs of glaciation, such as the global cooling initiated in the Oligocene. Extending from 35 Ma to the present period of Quaternary glaciations, this period is only the most recent of several prolonged, extensive glaciations ("megaglaciations"), each lasting several tens of Ma and separated by hundreds of Ma. Meglaciations also occurred during the Permian/Carboniferous, 320–260 Ma; the Neoproterozoic, 750–570 Ma; the Paleoproterozoic, 2400–2200 Ma; and possibly the Archean, 2990–2910 Ma (Crowell, 1999).

The Neoproterozoic megaglaciations are divided into an older phase, the Sturtian, and a younger phase, the Marinoan. The deposits associated with these glaciations are remarkable for (1) their widespread, low-latitude distribution (Harland, 1964; Evans, 2000); (2) the presence of banded iron formations (BIFs) which otherwise are common in the Earth record only prior to ~1.85 Ga (Klein and Beukes, 1993), probably because of the evolution of oxygen in the atmosphere (Holland, 1994); and (3) very unusual cap carbonate deposits that commonly overlie the glacial deposits with knife-sharp contacts (Hoffman et al., 1998). Associated varved deposits and fossil ice wedges are interpreted to result from very strong seasonal temperature variations over long time periods (Williams, 1993). Williams (1975, 1993, 2005) explains these features as the result of extremely high obliquity (>54°) during Earth's early history.

Kirschvink (1992) introduced a hypothesis for the Neoproterozoic glaciations that he named the "Snowball Earth Hypothesis" (SBEH). During the Neoproterozoic, continental land masses were concentrated at middle to low latitudes. Because land is more reflective of solar energy than is ocean water, the energy-absorbing character of Earth's tropics was reduced. If temperatures could be lowered in such a situation by additional effects on the "greenhouse," runaway cooling might be initiated by the ice-albedo effect. One possible mechanism for this is the heavy precipitation over relatively small, low-latitude continents, resulting in carbon dioxide removal from the atmosphere because of intense silicate weathering (Hoffman and Schrag, 2002). As the Earth entered its "icehouse" phase, a global freezing-over would isolate the oceans from the oxygenating effect of the atmosphere. The resulting anoxic conditions then lead to the concentration in seawater of ferrous iron from submarine hydrothermal sources (Kirschvink, 1992). However, as carbon dioxide was introduced by continuing volcanic emissions, atmospheric concentrations would build up, without the removal process of silicate weathering on the ice-covered continents. This eventually results, after several Ma, in a "supergreenhouse" that melts the global ice cover, reoxygenating the oceans so that they rapidly deposit their iron as the otherwise enigmatic BIFs (Kirschvink, 1992). Hoffman et al. (1998) added the explanation of the cap carbonates as the result of the newly exposed continental land surfaces experiencing especially intense weathering in the "supergreenhouse." The resulting immense flux of alkalinity (metal cations and bicarbonate) to the oceans would cover the cold-climate glacial deposits with the warm-climate carbonates.

The SBEH is criticized for its oversimplification of complex details of glaciological history and its presumed synchroneity of world-wide glaciations (Young, 2005). However, recent support for the cataclysmic terminations of the Marionoan and Sturtian glaciations is provided by the observation of sharp spikes of iridium in cap carbonates from the Eastern Congo craton (Bodiselitsch et al., 2005). The iridium is presumably introduced from extraterrestrial inputs to the global ice cover over several Ma. Rapid melting of the ice in the glacial-terminating "supergreenhouse" would then introduce the observed iridium spikes into the cap carbonates.

Schrag et al. (2002) propose that "snowball Earth" glaciations could have been initiated because the concentration of continental landmasses in the tropics would have been efficient sites for the burial of organic carbon. The associated oceanic anoxic conditions, generated by oceanic upwelling near these continents, would lead to methane production, concentrating the methane in gas hydrate reservoirs. The slow release of this methane would lead to the observed low values δC^{13} immediately prior to the glaciations. The enhanced greenhouse conditions from the methane would accelerate CO_2 removal from the atmosphere via silicate weathering on the tropical continents. Thus a methane greenhouse would gradually replace the carbon dioxide greenhouse. However, this situation is highly unstable because methane is rapidly destroyed by oxidation in Earth's atmosphere. When the gas hydrate reservoirs were exhausted, new methane could not be added to the atmosphere to replace what was lost, leading to the rapid collapse of the "greenhouse" and a runaway ice-albedo feedback that would take the planet to "icehouse" conditions, i.e., to a snowball glaciation.

There is considerable evidence suggesting that "snowball Earth" glaciations may have provided critical stimuli in regard to the evolution of life. The Neoproterozoic glaciations were followed by the Ediacara, the first large and diverse assemblage of animal fossils. The Paleoproterozoic event immediately precedes a phase of oxidative precipitation of iron and manganese that can be interpreted as a result of evolutionary branching in a superoxide enzyme in archaea and bacteria (Kirschvink et al., 2000).

"MEGAOUTFLO" EPISODES

The immense floods that initially fed Mars's hypothetical Oceanus Borealis could have been the triggers for hydroclimatic change through the release of radiatively active gases, including CO_2 and water vapor (Gulick et al., 1997). During the short-duration thermal episodes of cataclysmic outflow, a temporary cool-wet climate would prevail. Water that evaporated off Oceanus Borealis would transfer to uplands, including the Tharsis volcanoes and portions of the southern highlands, where precipitation as snow promoted the growth of glaciers, and rain contributed to valley development and lakes. However, this cool-wet climate was inherently unstable. Water from the evaporating surface-water bodies was lost to storage (1) in the highland glaciers and (2) as infiltration into the porous lithologies of the Martian surface. This latter effect, more than any lack of precipitation, is the likely cause of the limited upland dissection on Mars (Baker and Partridge, 1986). However, recent high-resolution data show that previous notions of limited upland dissection (e.g., Carr, 1996) were mislead by the available lower resolution imagery.

Progressive loss of carbon dioxide from the quasistable atmospheres of MEGAOUTFLO episodes was probably facilitated by

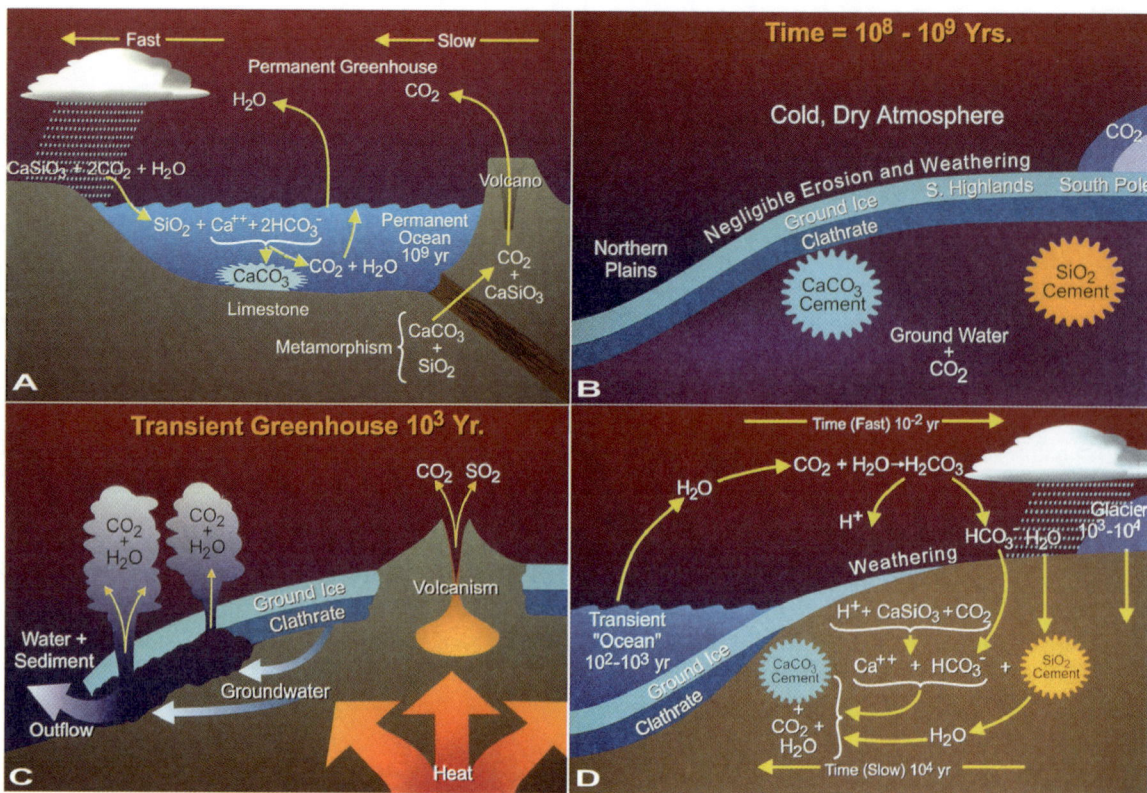

Figure 1. Schematics illustrating long-term cycling of water and carbon dioxide on Earth and Mars. (A) Earth's long-term CO_2 cycle. (B) North-south cross section of Mars's upper crust during the long-term quiescent phase of cold-dry surface conditions, where ground, containing carbon dioxide and perhaps methane, is confined beneath a cryosphere (icy permafrost) that is rich in gas hydrates. (C) MEGAOUTFLO event in which the crust (Fig. 1B) is disrupted by a short-term pulse for tectonic and volcanic activity, while huge outflow events deliver water and sediment to the northern plains and radiatively active gases to the atmosphere. (D) The result is a more Earth-like Mars (compare to Fig. 1A) in which a temporary greenhouse warming promotes active hydrological cycling through the atmosphere; however, this greenhouse would be rapidly terminated because of water storage as glaciers, infiltration of surface water into the porous lithologies of the Martian surface, and silicate weathering that would transfer bicarbonates to the subsurface with the infiltrating water; in centuries or millennia Mars would revert to the conditions illustrated in Figure 1B.

(1) dissolved gas in infiltrating acidic water and (2) silicate weathering producing bicarbonate that was carried into the subsurface by infiltration. Subsequent underground carbonate precipitation would then release some of the CO_2 back to the groundwater. However, the developing permafrost from a cooling climate would have trapped this gas in the subsurface. The permafrost zone would have developed as the surface greenhouse effect, promoted by a MEGAOUTFLO episode, went into decline because of the progressive loss of its gases to the subsurface by solution in infiltrating waters. As the permafrost extended downward into the stability field for CO_2 clathrate, this gas hydrate would then accumulate above the gas-charged groundwater. Thus the long-term reservoir for carbon on Mars is a sequestering underground in the forms of (1) clathrate (gas hydrate), (2) gas-charged groundwater, and (3) carbonate cements in crustal rocks. Only occasionally, and for relatively short duration, does carbon get transferred to the atmosphere, as greenhouse-promoting carbon dioxide, during the cataclysmic ocean-forming (MEGAOUTFLO) episodes. Oceanus Borealis (and the later smaller scale lacustrine episodes) does not last long enough in any individual episode for appreciable carbonate deposition. Moreover, the volcanism and water vapor associated with MEGAOUTFLO generate an acidic atmosphere (water, sulfur dioxide, carbon dioxide) that promotes the formation of salt crusts over the entire surface. These ideas seem to explain the lack of observed carbonates in spectra from the Thermal Emission Spectrometer (TES) instrument on Mars Global Surveyor. The short duration of the ocean-forming phases is also consistent with the very low degradation rates (Golombek and Bridges, 2000) for the Martian surface during the long period after heavy bombardment.

The recent discovery of small amounts of methane in the Martian atmosphere (Krasnopolsky et al., 2004) opens up an even more intriguing scenario for MEGAOUTFLO. The methane could be derived from a deep biosphere of methanogens in the Martian groundwater. Isolated below the ice-rich permafrost, this biosphere would produce the more effective greenhouse gas

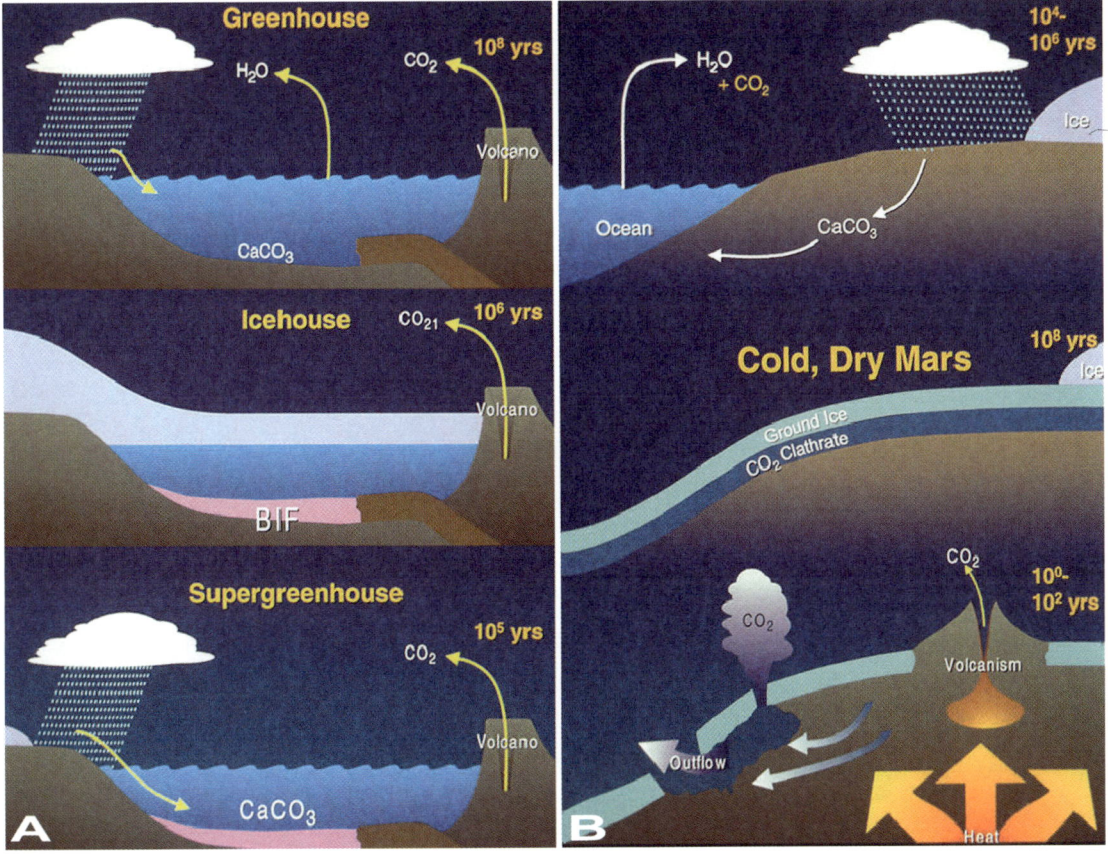

Figure 2. Schematic diagrams comparing the Snowball Earth Hypothesis (A) with the Mars MEGAOUTFLO hypothesis (B; MEGAOUTFLO diagrams correspond, respectively, from top to bottom to Figures 1D, 1B, and 1C). (A) Earth's stable state (upper left) has a persistent ocean in which carbon is precipitated from bicarbonate that is mobilized by the weathering of its landmasses (see also Fig. 1A); if perturbed to a runaway ice-albedo feedback, the ocean can be largely isolated from the atmosphere (middle left), achieving a so-called "snowball" state. The ocean will locally become anoxic, permitting the mobilization of iron and its eventual sequestering as banded iron formations (BIF); however, the continuing outgassing of carbon dioxide to the atmosphere will eventually create a supergreenhouse, melting the ice and producing the very intense weathering that brings cap carbonates to the oceans (lower left). Each of these phases has an equivalent in the MEGAOUTFLO scenario (right on "B").

methane, which would be stored in the ground ice as gas hydrate. Release of methane by MEGAOUTFLO episodes (one of which seems to have occurred in very recent geological time) would then have produced even more effective quasistable atmospheric warming than would be possible with carbon dioxide.

DISCUSSION

In a very broad sense, the Mars cataclysmic outflow events and associated cycles of MEGAOUTFLO can be considered to be broadly analogous to the terrestrial episodes of supercontinent assembly and breakup that are associated with megaglaciations. Continental positioning impacted Earth's long-term carbon cycling, perhaps triggering the spectacular "snowball" glaciations. Icehouse/greenhouse transitions occurred on both planets. Because of the continuous presence of an ocean on its surface, the megafloods of Earth seem to have been only effective at changing climate within an overall glacial epoch. For Mars, the MEGA-OUTFLO hypothesis envisions a central role for megaflooding to induce global climate change.

The stable state for Earth throughout its geological history is one of balance between (1) carbon dioxide input to the atmosphere by volcanism and (2) carbon dioxide removal by weathering of continental rocks and precipitation as marine carbonates (Fig. 1A). In contrast, the stable state for long periods of Mars history is carbon dioxide sequestration in groundwater and ground ices (Fig. 1B), including clathrates. During the Proterozoic megaglaciations, however, Earth switched to a cold-dry, Mars-like icehouse. The SBEH proposes that this occurs by freezing over of the global ocean. This extreme state is unstable, and it is terminated cataclysmically when carbon dioxide released from volcanism builds to a critical supergreenhouse level, resulting in melting of all surface ice and the very rapid drawdown of atmospheric CO_2 and associated precipitation of carbonates

(Fig. 2). Similarly, MEGAOUTFLO holds that Mars's stable cold-dry state is terminated by cataclysmic outburst flooding, injecting carbon dioxide (and perhaps other greenhouse gases) into the atmosphere (Fig. 1C). For short periods of time (perhaps thousands of years), Mars experienced an Earth-like water/carbon dioxide cycle (Fig. 1D).

ACKNOWLEDGMENTS

This chapter is Contribution No. 66 of the Arizona Laboratory for Paleohydrological and Hydroclimatological Analysis.

REFERENCES CITED

Baker, V.R., 1973, Paleohydrology and sedimentology of Lake Missoula flooding in eastern Washington: Geological Society of America Special Paper 144, 79 p.

Baker, V.R., ed., 1981, Catastrophic Flooding: The Origin of the Channeled Scabland: Stroudsburg, Pennsylvania, Dowden, Hutchinson and Ross, 360 p.

Baker, V.R., 1982, The Channels of Mars: Austin, Texas, University of Texas Press, 198 p.

Baker, V.R., 2001, Water and the Martian landscape: Nature, v. 412, p. 228–236, doi: 10.1038/35084172.

Baker, V.R., 2002, High-energy megafloods: Planetary settings and sedimentary dynamics, in Martini, I.P., Baker, V.R., and Garzon, G., eds., Flood and Megaflood Processes and Deposits: Recent and Ancient Examples: International Association of Sedimentologists, Special Paper No. 32, p. 3–15.

Baker, V.R., 2003, Planetary science: Icy Martian mysteries: Nature, v. 426, p. 779–780, doi: 10.1038/426779a.

Baker, V.R., 2004, A brief geological history of water on Mars, in Seckbach, J., ed., Origins: Genesis, evolution, and biodiversity of microbial life in the universe: Dordrecht, The Netherlands, Kluwer, p. 621–631.

Baker, V.R., 2005, Picturing a recently active Mars: Nature, v. 434, p. 280–283, doi: 10.1038/434280a.

Baker, V.R., and Milton, D.J., 1974, Erosion by catastrophic floods on Mars and Earth: Icarus, v. 23, p. 27–41, doi: 10.1016/0019-1035(74)90101-8.

Baker, V.R., and Partridge, J., 1986, Small Martian valleys: Pristine and degraded morphology: Journal of Geophysical Research, v. 91, p. 3561–3572, doi: 10.1029/JB091iB03p03561.

Baker, V.R., Strom, R.G., Gulick, V.C., Kargel, J.S., Komatsu, G., and Kale, V.S., 1991, Ancient oceans, ice sheets and hydrological cycle on Mars: Nature, v. 352, p. 589–594.

Baker, V.R., Carr, M.H., Gulick, V.C., Williams, C.R., and Marley, M.S., 1992, Channels and valley networks, in Kieffer, H.H., Jakosky, B., and Snyder, C., eds., Mars: Tucson, University of Arizona Press, p. 493–522.

Baker, V.R., Benito, G., and Rudoy, A.N., 1993, Paleohydrology of late Pleistocene superflooding, Altai Mountains, Siberia: Science, v. 259, p. 348–350, doi: 10.1126/science.259.5093.348.

Baker, V.R., Strom, R.G., Dohm, J.M., Gulick, V.C., Kargel, J.S., Komatsu, G., Ori, G.G., and Rice, J.W., Jr., 2000, Mars' Oceanus Borealis, Ancient Glaciers, and the MEGAOUTFLO Hypothesis, in Lunar and Planetary Science XXXI: Houston, Texas, Lunar and Planetary Institute, Abstract No. 1863.

Barber, D.C., Dyke, A., Hillaire-Marcel, C., Jennings, A.E., Andrews, J.T., Kerwin, M.W., Bilodeau, G., McNeely, R., Southon, J., Morehead, M.D., and Gagnon, J.M., 1999, Forcing of the cold event of 8200 years ago by catastrophic drainage of Laurentide lakes: Nature, v. 400, p. 344–348, doi: 10.1038/22504.

Benn, D.I., and Evans, D.J.A., 1998, Glaciers and Glaciation: London, Arnold, 734 p.

Berman, D.C., and Hartmann, W.K., 2002, Recent fluvial, volcanic, and tectonic activity on the Cerberus Plains of Mars: Icarus, v. 159, p. 1–17, doi: 10.1006/icar.2002.6920.

Björnsson, H., 2002, Subglacial lakes and jökulhlaups in Iceland: Global and Planetary Change, v. 35, p. 255–271, doi: 10.1016/S0921-8181(02)00130-3.

Blanchon, P., and Shaw, J., 1995, Reef drowning during the last deglaciation: Evidence for catastrophic sea-level rise and ice sheet collapse: Geology, v. 23, p. 4–8, doi: 10.1130/0091-7613(1995)023<0004:RDDTLD>2.3.CO;2.

Bodiselitsch, B., Koeberl, C., Master, S., and Reimold, W.U., 2005, Estimating duration and intensity of Neoproterozoic snowball glaciations from Ir anomalies: Science, v. 308, p. 239–242.

Bond, G.C., and Lotti, R., 1995, Iceberg discharges into the North Atlantic on millennial time-scales during the last glaciation: Science, v. 267, p. 1005–1010, doi: 10.1126/science.267.5200.1005.

Boynton, W.V., Feldman, W.C., Squyres, S.W., Prettyman, T.H., Brückner, J., Evans, L.G., Reedy, R.C., Starr, R., Arnold, J.R., Drake, D.M., Englert, P.A.J., Metzger, A.E., Mitrofanov, I., Trombka, J.I., d'Uston, C., Wänke, H., Gasnault, O., Hamara, D.K., Janes, D.M., Marcialis, R.L., Maurice, S., Mikheeva, I., Taylor, G.J., Tokar, R., and Shinohara, C., 2002, Distribution of Hydrogen in the near surface of Mars: Evidence for subsurface ice deposits: Science, v. 297, p. 81–85, doi: 10.1126/science.1073722.

Brennand, T.A., and Shaw, J., 1994, Tunnel channels and associated landforms: Their implications for ice sheet hydrology: Canadian Journal of Earth Sciences, v. 31, p. 502–522.

Broecker, W.S., Kennett, J.P., Flower, B.P., Teller, J.T., Trumbore, S., Bonani, G., and Wolfi, W., 1989, The routing of meltwater from the Laurentide ice-sheet during the Younger Dryas cold episode: Nature, v. 341, p. 318–321, doi: 10.1038/341318a0.

Budyko, M.I., 1966, Polar ice and climate, in Fletcher, G., ed., Proceedings of the Symposium on the Arctic Heat Budget and Atmospheric Circulation: Santa Monica, California, The Rand Corp., p. 3–21.

Budyko, M.I., 1969, The effect of solar radiation on the climate of the Earth: Tellus, v. 21, p. 611–619.

Burr, D.M., McEwen, A., and Sakimoto, S.E.H., 2002, Recent aqueous floods from the Cerberus Fossae, Mars: Geophysical Research Letters, v. 29, doi: 10.1029/2001GL013345.

Cabrol, N.A., and Grin, E.A., 1999, Distribution, classification, and ages of martian impact crater lakes: Icarus, v. 142, p. 160–172, doi: 10.1006/icar.1999.6191.

Cabrol, N.A., and Grin, E.A., 2001, The evolution of lacustrine environments on Mars: Is Mars only hydrologically dormant?: Icarus, v. 149, p. 291–328, doi: 10.1006/icar.2000.6530.

Carr, M.H., 1979, Formation of Martian flood features by release of water from confined aquifers: Journal of Geophysical Research, v. 84, p. 2995–3007, doi: 10.1029/JB084iB06p02995.

Carr, M.H., 1996, Water on Mars: Oxford, Oxford University Press, 229 p.

Carr, M.H., 2000, Martian oceans, valleys and climate: Astronomy and Geophysics, v. 41, p. 3.21–3.26.

Chapman, M.G., 1994, Evidence, age, and thickness of a frozen paleolake in Utopia Planitia, Mars: Icarus, v. 109, p. 393–406, doi: 10.1006/icar.1994.1102.

Chapman, M.G., Gudmundsson, M.T., Russell, A.J., and Hare, T.M., 2003, Possible Juventae Chasma sub-ice volcanic eruptions and Maja Valles ice outburst floods, Mars: Implications of MGS crater densities, geomorphology, and topography: Journal of Geophysical Research, v. 108, doi: 10.1029/2002JE002009.

Clarke, G., Leverington, D., Teller, J., and Dyke, A., 2004, Paleohydraulics of the last outburst flood from glacial Lake Agassiz and the 8200 BP cold event: Quaternary Science Reviews, v. 23, p. 389–407, doi: 10.1016/j.quascirev.2003.06.004.

Clarke, G.K.C., Leverington, D.W., Teller, J.T., Dyke, A.S., and Marshall, S.J., 2005, Fresh arguments against the Shaw megaflood hypothesis: A reply to comments by David Sharpe on "Paleohydraulics of the last outburst flood from glacial Lake Agassiz and the 8200 BP cold event": Quaternary Science Reviews, v. 24, p. 1533–1541, doi: 10.1016/j.quascirev.2004.12.003.

Clifford, S.M., 1993, A model for the hydrologic and climatic behavior of water on Mars: Journal of Geophysical Research, v. 98, p. 10,973–11,016, doi: 10.1029/93JE00225.

Clifford, S.M., and Parker, T.J., 2001, The evolution of the martian hydrosphere: Implications for the fate of a primordial ocean and the current state of the northern plains: Icarus, v. 154, p. 40–79, doi: 10.1006/icar.2001.6671.

Crowell, J.C., 1999, Pre-Mesozoic ice ages: Their bearing on understanding the climate system: Geological Society of America Memoir, v. 192, 106 p.

Denton, G.H., and Sugden, D.E., 2005, Meltwater features that suggest Miocene ice-sheet overriding of the Transantarctic Mountains in Victoria Land, Antarctica: Geografiska Annaler, v. 87A, p. 1–19.

Donnelly, J.P., Driscoll, N.W., Uchupi, E., Kelgwin, L.D., Schwab, W.C., Theiler, E.R., and Swift, S.A., 2005, Catastrophic meltwater discharge down the Hudson Valley: A potential trigger for the Intra-Allerod cold period: Geology, v. 33, p. 89–92, doi: 10.1130/G21043.1.

Dugdale, R.C., Lyle, M., Wilkerson, F.P., Chai, F., Barber, R.T., and Peng, T.H., 2004, Influence of equatorial diatom processes on Si deposition and atmospheric CO_2 cycles at glacial/interglacial timescales: Paleoceanography, v. 19, doi: 10.1029/2003PA000929.

Erikson, E., 1968, Air-ocean-icecap interactions in relation to climate fluctuations and glaciation Cycles: Meteorological Monographs, v. 8, p. 68–92.

Evans, D.A., 2000, Stratigraphic, geochronological, and paleomagnetic constraints upon the Neoproterozoic climatic paradox: American Journal of Science, v. 300, p. 347–433, doi: 10.2475/ajs.300.5.347.

Fairbanks, R.G., 1989, A 17,000-year glacio-eustatic sea level record: Influence of glacial melting rates on the Younger Dryas event and deep ocean circulation: Nature, v. 342, p. 637–642, doi: 10.1038/342637a0.

Fisher, T.G., and Shaw, J., 1993, A depositional model for Rogen moraine, with examples from the Avalon Peninsula, Newfoundland, Canada: Canadian Journal of Earth Sciences, v. 29, p. 669–686.

Golombek, M.P., and Bridges, N.T., 2000, Erosion rates on Mars and implications for climate change: Constraints from the Pathfinder landing site: Journal of Geophysical Research, v. 105, p. 1841–1853, doi: 10.1029/1999JE001043.

Gough, D.O., 1981, Solar interior structure and luminosity variations: Solar Physics, v. 74, p. 21–34, doi: 10.1007/BF00151270.

Gross, M.G., 1987, Oceanography: A view of the Earth: Englewood Cliffs, Prentice-Hall, 406 p.

Grosswald, M.G., 1980, Late Weichselian Ice Sheet of northern Eurasia: Quaternary Research, v. 13, p. 1–32, doi: 10.1016/0033-5894(80)90080-0.

Grosswald, M.G., 1999, Cataclysmic Megafloods in Eurasia and the Polar Ice Sheets: Moscow, Scientific World (in Russian), 120 p.

Gulick, V.C., Tyler, D., McKay, C.P., and Haberle, R.M., 1997, Episodic ocean-induced CO_2 pulses on Mars: Implications for fluvial valley formation: Icarus, v. 130, p. 68–86, doi: 10.1006/icar.1997.5802.

Harland, W.B., 1964, Evidence or late Precambrian glaciation and its significance, in Nairn, A.E.M., ed., Problems in paleoclimatology: London, Wiley, p. 119–149.

Hartmann, W.K., and Berman, D.C., 2000, Elysium Planitia lava flows: Crater count chronology and geological implications: Journal of Geophysical Research, v. 105, p. 15,011–15,025, doi: 10.1029/1999JE001189.

Head, J.W., Hiesinger, H., Ivanov, M.A., Kreslavsky, M.A., Pratt, S., and Thomson, B.J., 1999, Possible ancient oceans on Mars: Evidence from Mars Orbiter Laser Altimeter data: Science, v. 286, p. 2134–2137, doi: 10.1126/science.286.5447.2134.

Head, J.W., Kreslavsky, M.A., and Pratt, S., 2002, Northern lowlands of Mars: Evidence for widespread volcanic flooding and tectonic deformation of the Hesperian Period: Journal of Geophysical Research, v. 107, doi: 10.1029/2000JE001445.

Head, J.W., Mustard, J.F., Kreslavsky, M.A., Milliken, R.E., and Marchant, D.R., 2003, Recent ice ages on Mars: Nature, v. 426, p. 797–802, doi: 10.1038/nature02114.

Herget, J., 2005, Reconstruction of Pleistocene Ice-Dammed Lake Outburst Floods in the Altai Mountains, Siberia: Geological Society of America Special Paper 386, 118 p.

Hesse, R., Rashid, H., and Khodabkhsk, S., 2004, Fine-grained sediment lofting from meltwater-generated turbidity currents during Heinrich events: Geology, v. 32, p. 449–452, doi: 10.1130/G20136.1.

Hoffman, N., 2000, White Mars: A new model for Mars' surface and atmosphere based on CO_2: Icarus, v. 146, p. 326–342, doi: 10.1006/icar.2000.6398.

Hoffman, P.F., and Schrag, D.P., 2002, The snowball Earth hypothesis: Testing the limits of global change: Terra Nova, v. 14, p. 129–155, doi: 10.1046/j.1365-3121.2002.00408.x.

Hoffman, P.F., Kaufman, A.J., Halverson, G.P., and Schrag, D.P., 1998, A Neoproterozoic snowball Earth: Science, v. 281, p. 1342–1346, doi: 10.1126/science.281.5381.1342.

Holland, H.D., 1994, Early Proterozoic atmospheric change, in Bengston, S., ed., Early life on Earth: Nobel Symposium No. 84, New York, Columbia University Press, p. 237–244.

Irwin, R.P., III, Howard, A.D., and Maxwell, T.A., 2004, Geomorphology of Ma'adim Vallis, Mars, and associated paleolake basins: Journal of Geophysical Research, v. 109, doi: 10.1029/2004JE002287.

Ivanov, M.A., and Head, J.W., III, 2001, Chryse Planitia, Mars: Topographic configuration, outflow channel continuity and sequence, and tests for hypothesized ancient bodies of water using Mars Orbiter Laser Altimeter (MOLA) data: Journal of Geophysical Research, v. 106, p. 3275–3295, doi: 10.1029/2000JE001257.

Kargel, J.S., 2004, Mars: A Warmer Wetter Planet: Berlin, Springer, 557 p.

Kargel, J.S., Tanaka, K.A., Baker, V.R., and Komatsu, G., 2000, Formation and Dissociation of Clathrate Hydrates on Mars: Polar Caps, Northern Plains, and Highlands Crust, in Lunar and Planetary Science XXXI: Houston, Texas, Lunar and Planetary Institute, Abstract No. 1891.

Kehew, A.E., and Teller, J.T., 1994, History of the late glacial runoff along the southwestern margin of the Laurentide Ice Sheet: Quaternary Science Reviews, v. 13, p. 859–877, doi: 10.1016/0277-3791(94)90006-X.

Kennett, J.P., Cannariato, K.G., Hendy, I.L., and Behl, R.J., 2000, Carbon isotopic evidence for Methane hydrate instability during Quaternary interstadials: Science, v. 288, p. 128–133, doi: 10.1126/science.288.5463.128.

Kerr, R.A., 1993, An "Outrageous Hypothesis" for Mars: Episodic Oceans: Science, v. 259, p. 910–911.

Kirschvink, J.L., 1992, Late Proterozoic low-latitude global glaciation: The snowball Earth, in Schopf, J.W., and Klein, C., eds., The Proterozoic biosphere: A multidisciplinary study: Cambridge, UK, Cambridge University Press, p. 51–52.

Kirschvink, J.L., Gaidos, E.J., Bertanie, L.E., Beukes, N.J., Gutzmer, J., Maepa, L.N., and Steinberger, R.E., 2000, Paleoproterozoic snowball Earth: Extreme climatic and geochemical global change and its biological consequences: Proceedings of the National Academy of Sciences of the United States of America, v. 97, no. 4, p. 1400–1405, doi: 10.1073/pnas.97.4.1400.

Klein, C., and Beukes, N.J., 1993, Sedimentology and geochemistry of the glaciogenic late Proterozoic Rapitan iron-formation in Canada: Economic Geology and the Bulletin of the Society of Economic Geologists, v. 84, p. 1733–1774.

Komar, P., 1979, Comparisons of the hydraulics of water flows in the Martian outflow channels with flows of similar scale on Earth: Icarus, v. 37, p. 156–181, doi: 10.1016/0019-1035(79)90123-4.

Komar, P.D., 1980, Modes of sediment transport in channelized flows with ramifications to the erosion of Martian outflow channels: Icarus, v. 42, p. 317–329, doi: 10.1016/0019-1035(80)90097-4.

Komatsu, G., Kargel, J.S., Baker, V.R., Strom, R.G., Ori, G.G., Mosangini, G., and Tanaka, K.L., 2000, A chaotic terrain formation hypothesis: Explosive outgas and outflow by dissociation of clathrate on Mars, in Lunar and Planetary Science XXXI: Houston, Texas, Lunar and Planetary Institute, Abstract No. 1434.

Kor, P., Shaw, J., and Sharpe, D.R., 1991, Erosion of bedrock by subglacial meltwater, Georgian Bay, Ontario, a regional review: Canadian Journal of Earth Sciences, v. 28, p. 623–642.

Kourafalou, V.H., Oey, L.Y., Wang, J.D., and Lee, T.N., 1996, The fate of river discharge on the Continental shelf. 1: Modeling the river plume and the inner shelf coastal current: Journal of Geophysical Research, v. 101, p. 3415–3434, doi: 10.1029/95JC03024.

Krasnopolsky, V.A., Maillard, J.P., and Owen, T.C., 2004, Detection of methane in the Martian Atmosphere: Evidence for life?: Icarus, v. 172, p. 537–547, doi: 10.1016/j.icarus.2004.07.004.

Kuzmin, R.O., Bobina, N.N., Zabalueva, E.V., and Shashkina, V.P., 1988, Structural inhomogeneities of the Martian cryolithosphere: Solar System Research, v. 22, p. 195–212.

Lucchitta, B.K., Ferguson, H.M., and Summers, C., 1986, Sedimentary deposits in the northern lowland plains, Mars: Journal of Geophysical Research, v. 91, p. E166–E174, doi: 10.1029/JB091iB13p0E166.

Malin, M.C., and Edgett, K.S., 1999, Oceans or seas in the Martian northern lowlands: High-resolution imaging tests of proposed coastlines: Geophysical Research Letters, v. 26, p. 3049–3052, doi: 10.1029/1999GL002342.

Malin, M.C., and Edgett, K.S., 2001, Global Surveyor Orbiter Camera: Inter-planetary cruise through primary mission: Journal of Geophysical Research, v. 106, p. 23,429–23,570, doi: 10.1029/2000JE001455.

Masursky, H., Boyce, J.M., Dial, A.L., Schaber, G.C., and Strobell, M.E., 1977, Classification and time of formation of Martian channels based on Viking data: Journal of Geophysical Research, v. 82, p. 4016–4038, doi: 10.1029/JS082i028p04016.

Max, M.D., and Clifford, S.M., 2001, Initiation of Martian outflow channels: Related to the dissociation of gas hydrate?: Geophysical Research Letters, v. 28, p. 1787–1790, doi: 10.1029/2000GL011606.

Milton, D., 1974, Carbon dioxide hydrate and floods on Mars: Science, v. 183, p. 654–656, doi: 10.1126/science.183.4125.654.

Mouginis-Mark, P.J., 1990, Recent water release in the Tharsis region of Mars: Icarus, v. 84, p. 362–373, doi: 10.1016/0019-1035(90)90044-A.

Mulder, T., Syvitski, J.P.M., Migeon, S., Faugeres, J.C., and Savoye, B., 2003, Marine hyperpycnal flows: Initiation behavior and related deposits: A review: Marine and Petroleum Geology, v. 20, p. 861–882, doi: 10.1016/j.marpetgeo.2003.01.003.

Munro, M., and Shaw, J., 1997, Erosional origin of hummocky terrain, south central Alberta, Canada: Geology, v. 25, p. 1027–1030, doi: 10.1130/0091-7613(1997)025<1027:EOOHTI>2.3.CO;2.

Normark, W.R., and Reid, J.A., 2003, Extensive deposits on the Pacific Plate from Late Pleistocene North-American glacial lake bursts: The Journal of Geology, v. 111, p. 617–637, doi: 10.1086/378334.

Nummedal, D., and Prior, D., 1981, Generation of Martian chaos and channels by debris Flows: Icarus, v. 45, p. 77–86, doi: 10.1016/0019-1035(81)90007-5.

O'Connor, J.E., 1993, Hydrology, Hydraulics and Sediment Transport of Pleistocene Lake Bonneville Flooding on the Snake River, Idaho: Geological Society of America Special Paper 274, 83 p.

O'Connor, J.E., and Baker, V.R., 1992, Magnitudes and implications of peak discharges from glacial lake Missoula: Geological Society of America Bulletin, v. 104, p. 267–279, doi: 10.1130/0016-7606(1992)104<0267:MAIOPD>2.3.CO;2.

Paige, D.A., 2005, Ancient Mars: Wet in many places: Science, v. 307, p. 1575–1576, doi: 10.1126/science.1110530.

Parker, T.J., Saunders, R.S., and Schneeberger, D.M., 1989, Transitional morphology in the west Deuteronilus Mensae region of Mars: Implications for modification of the lowland/upland boundary: Icarus, v. 82, p. 111–145, doi: 10.1016/0019-1035(89)90027-4.

Parker, T.J., Gorsline, D.S., Saunders, R.S., Pieri, D., and Schneeberger, D.M., 1993, Coastal geomorphology of the Martian northern plains: Journal of Geophysical Research, v. 98, p. 11,061–11,078, doi: 10.1029/93JE00618.

Pierrehumbert, R.T., 2004, High levels of atmospheric carbon dioxide necessary for the Termination of global glaciation: Nature, v. 429, p. 646–649, doi: 10.1038/nature02640.

Robinson, M., and Tanaka, K., 1990, Magnitude of a catastrophic flood event at Kasei Vallis, Mars: Geology, v. 18, p. 902–905, doi: 10.1130/0091-7613(1990)018<0902:MOACFE>2.3.CO;2.

Ryan, W.B.F., Major, C.O., Lericolais, G., and Goldstein, S.L., 2003, Catastrophic flooding of the Black Sea: Annual Review of Earth and Planetary Sciences, v. 31, p. 525–554, doi: 10.1146/annurev.earth.31.100901.141249.

Schrag, D.P., Berner, R.A., Hoffman, P.F., and Halverson, G.P., 2002, On the initiation of a snow-ball Earth: Geochemistry, Geophysics, Geosystems, v. 3, 21 p., doi: 10.1029/2001GC000219.

Scott, D.H., Dohm, J.M., and Rice, J.W., 1995, Map of Mars showing channels and possible paleolakes: U.S. Geological Survey Miscellaneous Investigation Series Map I-2461.

Sellers, W.D., 1969, A global climate model based on the energy balance of the Earth-Atmosphere system: Journal of Applied Meteorology, v. 8, p. 392–400, doi: 10.1175/1520-0450(1969)008<0392:AGCMBO>2.0.CO;2.

Shaw, J., 1983, Drumlin formation related to inverted meltwater erosion marks: Journal of Glaciology, v. 29, p. 461–479.

Shaw, J., 1988, Subglacial erosional marks, Wilton Creek, Ontario: Canadian Journal of Earth Sciences, v. 25, p. 1256–1267.

Shaw, J., 1996, A meltwater model for Laurentide subglacial landscapes, in McCann, S.B., and Ford, D.C., eds., Geomorphology sans Frontieres: New York, Wiley, p. 182–226.

Shaw, J., 2002, The meltwater hypothesis for subglacial landforms: Quaternary Science Reviews, v. 90, p. 5–22.

Shaw, J., Kvill, D., and Rains, B., 1989, Drumlins and catastrophic subglacial floods: Sedimentary Geology, v. 62, p. 177–202, doi: 10.1016/0037-0738(89)90114-0.

Shoemaker, E.M., 1995, On the meltwater genesis of drumlins: Boreas, v. 24, p. 3–10.

Siegert, M.J., 2005, Lakes beneath the ice sheet: The occurrence, analysis, and future exploration of Lake Vostok and other Antarctic subglacial lakes: Annual Review of Earth and Planetary Sciences, v. 33, p. 215–246, doi: 10.1146/annurev.earth.33.092203.122725.

Sjogren, D., and Rains, R.B., 1995, Glaciofluvial erosional morphology and sediments of the Coronation-Spondin Scabland, east-central Alberta: Canadian Journal of Earth Sciences, v. 32, p. 565–578.

Smith, A.J., 1985, A catastrophic origin for the paleovalley system of the English Channel: Marine Geology, v. 64, p. 65–75, doi: 10.1016/0025-3227(85)90160-4.

Stuut, J.B., Crosta, X., van der Borg, K., and Schneider, R., 2004, Relationship between Antarctic sea ice and SW African climate during the late Quaternary: Geology, v. 32, p. 909–912, doi: 10.1130/G20709.1.

Sugden, D.E., and Denton, G.H., 2004, Cenozoic landscape evolution of the Convoy Range to Mackay Glacier area, Transantarctic Mountains: Onshore to offshore synthesis: Geological Society of America Bulletin, v. 116, p. 840–857, doi: 10.1130/B25356.1.

Tanaka, K.L., and Chapman, M.C., 1990, The relation of catastrophic flooding of Mangala Vallis, Mars, to faulting of Memnonia Fossae and Tharsis volcanism: Journal of Geophysical Research, v. 95, p. 14,315–14,323, doi: 10.1029/JB095iB09p14315.

Tanaka, K.L., Banerdt, W.B., Kargel, J.S., and Hoffman, N., 2001, Huge CO_2-charged debris-flow deposit and tectonic sagging in the northern plains of Mars: Geology, v. 29, p. 427–430, doi: 10.1130/0091-7613(2001)029<0427:HCCDFD>2.0.CO;2.

Tanaka, K.L., Skinner, J.A., Jr., and Hare, T.M., 2005, Geologic map of the northern plains of Mars: U.S. Geological Survey Miscellaneous Investigations Series Map I-2888.

Teller, J.T., Leverington, D.W., and Mann, J.D., 2002, Freshwater outbursts to the oceans from glacial Lake Agassiz and their role in climate change during the last deglaciation: Quaternary Science Reviews, v. 21, p. 879–887, doi: 10.1016/S0277-3791(01)00145-7.

Walder, J.S., 1994, Comments on "Subglacial" floods and the origin of low-relief ice-sheet lobes: Journal of Glaciology, v. 40, p. 199–200.

Walker, J.C.G., Hays, P.B., and Kasting, J.F., 1981, A negative feedback mechanism for the long-term stabilization of Earth's surface temperature: Journal of Geophysical Research, v. 86, p. 9776–9782, doi: 10.1029/JC086iC10p09776.

Williams, G.E., 1975, Late Precambrian glacial climate and the Earth's obliquity: Geological Magazine, v. 112, p. 441–465.

Williams, G.E., 1993, History of the Earth's obliquity: Earth-Science Reviews, v. 34, p. 1–45, doi: 10.1016/0012-8252(93)90004-Q.

Williams, G.E., 2005, The paradox of Proterozoic glaciomarine deposition, open seas and strong seasonality near the palaeo-equator: Global implications, in Eriksson, P.G., Altermann, W., Nelson, D.R., Mueller, W.U., and Catuneanu, O., eds., The Precambrian Earth: Tempos and Events: Amsterdam, Elsevier, p. 448–459.

Young, G.M., 2005, Earth's two great Precambrian glaciations: Aftermath of the "Snowball Earth" hypothesis, in Eriksson, P.G., Altermann, W., Nelson, D.R., Mueller, W.U., and Catuneanu, O., eds., The Precambrian Earth: Tempos and Events: Amsterdam, Elsevier, p. 440–448.

Zimbelman, J.R., Craddock, R.A., Greeley, R., and Kuzmin, R.O., 1992, Volatile history of Mangala Vallis, Mars: Journal of Geophysical Research, v. 97, p. 18,309–18,317.

MANUSCRIPT ACCEPTED BY THE SOCIETY 3 NOVEMBER 2008

Effects of megascale eruptions on Earth and Mars

Thorvaldur Thordarson*
School of GeoSciences, University of Edinburgh, Grant Institute, The King's Buildings, West Mains Road, Edinburgh, EH9 3JW, UK

Michael Rampino*
Earth & Environmental Science Program, New York University, 100 Washington Square East, Room 1009, New York, New York 10003, USA

Laszlo P. Keszthelyi*
Astrogeology Team, U.S. Geological Survey, 2255 N. Gemini Drive, Flagstaff, Arizona 86001, USA

Stephen Self*[†]
Volcano Dynamics Group, Department of Earth Sciences, The Open University, Milton Keynes, MK7 6AA, UK

ABSTRACT

Volcanic features are common on geologically active earthlike planets. Megascale or "super" eruptions involving >1000 Gt of magma have occurred on both Earth and Mars in the geologically recent past, introducing prodigious volumes of ash and volcanic gases into the atmosphere. Here we discuss felsic (explosive) and mafic (flood lava) supereruptions and their potential atmospheric and environmental effects on both planets. On Earth, felsic supereruptions recur on average about every 100–200,000 years and our present knowledge of the 73.5 ka Toba eruption implies that such events can have the potential to be catastrophic to human civilization. A future eruption of this type may require an unprecedented response from humankind to assure the continuation of civilization as we know it. Mafic supereruptions have resulted in atmospheric injection of volcanic gases (especially SO_2) and may have played a part in punctuating the history of life on Earth. The contrast between the more sustained effects of flood basalt eruptions (decades to centuries) and the near-instantaneous effects of large impacts (months to years) is worthy of more detailed study than has been completed to date. Products of mafic supereruptions, significantly larger than known from the geologic record on Earth, are well preserved on Mars. The volatile emissions from these eruptions most likely had global dispersal, but the effects may not have been outside what Mars endures even in the absence of volcanic eruptions. This is testament to the extreme variability of the current Martian atmosphere: situations that would be considered catastrophic on Earth are the norm on Mars.

---

*Emails: Thordarson: thor.thordarson@ed.ac.uk; Rampino: michael.rampino@nyu.edu; Keszthelyi: laz@usgs.gov; Self: Stephen.Self@open.ac.uk.
[†]Present address: US Nuclear Regulatory Commission, MS EBB-02-02, Washington, DC 20555-0001, USA.

Thordarson, T., Rampino, M., Keszthelyi, L.P., and Self, S., 2009, Effects of megascale eruptions on Earth and Mars, *in* Chapman, M.G., and Keszthelyi, L.P., eds., Preservation of random megascale events on Mars and Earth: Influence on geologic history: Geological Society of America Special Paper 453, p. 37–53, doi: 10.1130/2009.453(04). For permission to copy, contact editing@geosociety.org. ©2009 The Geological Society of America. All rights reserved.

INTRODUCTION

Products of volcanic eruptions are a common feature of geologically active earthlike planets. While Earth has more frequent eruptions than Mars, megascale eruptions took place on both bodies in the geologically recent past. These "super" eruptions have introduced prodigious volumes of ash and volcanic gases into the atmosphere. Sulfur that is converted into H_2SO_4 aerosols has a particularly dramatic effect on climate. In this chapter, we (1) evaluate current knowledge of the different types of large eruptions on Earth and Mars, (2) discuss the known effects on Earth, and (3) provide some improved estimates useful for future studies on volcanic effects on the Martian atmosphere.

When discussing megascale geologic events, it is convenient to use a logarithmic scale. Such a scale has been widely used for terrestrial explosive eruptions. Newhall and Self (1982) introduced the Volcanic Explosivity Index (VEI), based mainly on the volume of the erupted products and the height of the volcanic eruption column. VEIs range from VEI 0 (for strictly nonexplosive eruptions) to VEI 8 (for explosive eruptions producing ~1000 km³ bulk volume of tephra). Pyle (1995) later adapted the VEI to a logarithmic magnitude (M) scale, using the mass of the erupted magma such that $M = \log_{10}$ (erupted mass in kg) – 7.0. A density of 2400 kg/m³ is typical of both compacted felsic ash deposits and vesicular mafic lavas.

Before discussing the megascale eruptions, it is important to set a baseline from historical eruptions. One of the largest historical eruptions, the 1815 A.D. eruption of Tambora Volcano in Indonesia (VEI 7, M7), produced ~100 Mt of submicron stratospheric aerosols and dust (Self et al., 2004). Tambora is associated with a Northern Hemisphere climate cooling of ~1 °C (Stothers, 1983) (Table 1). The eruption was followed by unusually cold and wet conditions in 1816, which became known as the "year without a summer" in Europe and eastern North America. The Tambora event seems to have been the major factor in precipitating the last great subsistence crisis in Western Europe (Post, 1978; Stothers, 1983; Harington, 1992).

There have also been several large-volume basaltic eruptions in historical times (Simkin and Siebert, 1994). Although these largely effusive eruptions are VEI 4, volumetrically they are M6. Despite the relatively gentle nature of these eruptions, they introduced massive volumes of aerosols into the lower stratosphere (Stothers et al., 1986; Thordarson and Self, 1996; Thordarson et al., 1996; 2001, 2003a; Self et al., 2006) and affected regional and perhaps global climate (Carracedo et al., 1992; Thordarson and Self, 1993, 2003; Fiacco, et al., 1994; Stothers, 1996, 1998; Demaree et al., 1998; Grattan and Brayshay, 1995; Oman et al., 2006a, 2006b). The best studied example is the Laki (Skaftáreldar) eruption of 1783–1784 A.D. (Iceland), which produced ~14.7 km³ of lava flows and ~0.4 km³ (dense rock equivalent) of tehra (Thordarson and Self, 1993; Thordarson et al., 2003b). The resulting "haze famine" in Iceland, related to crop failures and livestock death from volcanic pollution (from SO_2, F, Cl, etc.), led to the death of 21% of the Icelandic population (Hálfdánarson, 1984). The 934–40 A.D. Eldgjá eruption may have had a similar serious effect on early Icelandic colonization (Stothers, 1998; Larsen, 2000; Thordarson et al., 2001).

In this chapter, we consider eruptions with VEI (or M) ≥8 to be "megascale" or "super" eruptions (e.g., Self et al., 2006). Others used slightly different thresholds for the term "supereruption" (e.g., Sparks et al., 2005). On Earth, there are two distinct types of supereruptions: (1) explosive caldera-forming events involving highly felsic magmas and (2) mafic flood lava eruptions. On Mars, there is clear evidence of flood lavas and intriguing hints of large explosive events—but no evidence of large felsic eruptions. Given our focus on processes that affected both Earth and Mars, we discuss the mafic supereruptions in more detail. However, as we explain next, the felsic supereruptions are of more direct concern for humanity.

FELSIC EXPLOSIVE SUPERERUPTIONS ON EARTH

The greatest volcanic eruption in the last few hundred-thousand years was the >M8 Toba (Sumatra) event of ~73.5 Ka (Chesner et al., 1991; Rampino and Self, 1992, 1993). This event produced at least 2800 km³ of lava (pyroclastic flow deposits, pumice fall, and ash), and is estimated to have created 1–10 Gt of stratospheric dust and sulfuric acid aerosols (Chesner et al., 1991; Rampino and Self, 1992; Zielinski et al., 1996a). Extrapolation of the data of Pyle et al. (1996) to >M8 eruptions gives ~1 Gt of SO_2 release, which would be converted to aerosols in the stratosphere. The Toba aerosols may have persisted for up to 6 years in the upper atmosphere (Rampino and Self, 1992, 1993; Zielinski et al., 1996a). By scaling up from smaller eruptions and numerical climate models, a stratospheric aerosol loading of 1 Gt is predicted to have caused a "volcanic winter," with a global cooling of 3–5 °C for several years, and regional cooling up to 15 °C (Rampino et al., 1988; Rampino and Self, 1992, 1993) (Fig. 1;

TABLE 1. SOME VOLCANIC ERUPTIONS AND THEIR ATMOSPHERIC EFFECTS

Volcano	Date	Stratospheric Aerosols (Gt)*	Cooling (°C)
St Helens	May 1980	0.0003	<0.1
Agung	March/May 1963	0.020	0.3
El Chichón	March/April 1982	0.014	0.3
Krakatau	August 1883	0.044	0.3
Pinatubo	June 1991	0.030	0.4
Tambora	April 1815	0.2	0.8
??	AD 1258	0.4–0.8	?
??	AD 536	0.3	?
Laki	June 1783–February 1784	0.2	1.0?
Eldgjá	AD 934	0.45	?
Roza (CRBG)	~15 Ma	9.6	?
Toba	~73.5 Ka	2.2–4.4	3 to 5?

Note: References: Rampino and Self (1984); Stothers (1983, 1984, 1996, 1998, 2000); Zielinski et al. (1996b); Thordarson et al. (1996, 2001); Thordarson and Self (1996, 2003).

*Aerosol mass is equal to 1.33 times H_2SO_4 mass.

Table 1). Such a cooling is estimated to have drastically affected tropical and temperate vegetation and ecosystems (Rampino and Ambrose, 2000). Aboveground tropical vegetation could have been killed by sudden hard freezes, and ~50% die-off of temperate forests is predicted from hard freezes during the growing season (Rampino and Ambrose, 2000; Sagan and Turco, 1990).

This probable climatic and ecologic disaster may have greatly impacted humans. Evidence from human genetic studies has suggested a severe human population bottleneck—a near extinction—with reductions to a total population as small as a few thousand, at a time just prior to ~60 ka (Harpending et al., 1993; Ambrose, 1998). This is roughly the same interval as the great Toba eruption, and a plausible cause and effect relationship with Toba has been proposed (Ambrose, 1998).

Frequency of Felsic Supereruptions

How frequent are Toba-scale (>M8) eruptions? A simple extrapolation of the pattern of recurrence times of smaller eruptions over the last 10 Ka leads to the prediction of recurrence times for >M8 eruptions of only a few thousand years (De la Cruz Reyna, 1991). This latter estimate of frequency for >M8 eruptions seems much too high (e.g., no Holocene >M8 eruptions have been recorded), and Simkin and Siebert (1994) argued that a simple extrapolation from smaller eruptions is not valid. A recent assessment by Self (2006) indicates supereruption frequency of roughly one every 100–200,000 years (Fig. 2). Such average periodicities are not that helpful because a) the eruptions appear to be clustered in time, and b) we have such an incomplete record.

While this 100–200,000 years recurrence interval is a useful metric, to obtain a better sense of the probability of a >M8 eruption in a given time period one must also examine the distribution of eruptions in time. The distribution of historical eruptions over the last few hundred years appears to be largely random (i.e., Poissonian) (Ho, 1991; Godano and Civetta, 1996). On longer time scales, however, there are hints that the frequency of eruptions may follow global climate cycles. A number of studies have found that large eruptions over the past few million years show cycles in the Milankovitch frequency band (e.g., ~23,000 to ~100,000 years; Paterne et al., 1990; Paterne and Guichard, 1993; Glazner et al., 1999). The link between enhanced volcanic activity and changes in climate is further strengthened by studies of Holocene volcanism (e.g., Hardarson and Fitton, 1991; Sigvaldason et al., 1992; Zielinski et al., 1994, 1996b; McGuire et al., 1997; Glazner et al., 1999; Maclennan et al., 2002), which indicate that an increase in magma production and eruption frequency in certain volcanic regions correlates with changes in climate proxies such as rise or fall of sea level or glacier advance and retreat. Several mechanisms involving loading and unloading of magma chambers by fluctuating ice sheets and sea levels have been proposed (e.g., Rampino et al., 1979; McGuire et al., 1997).

Evidence for temporal clustering of major explosive eruptions also comes from the GISP2 ice core, where several periods show groups of large sulfate peaks. One such interval (~22,000–

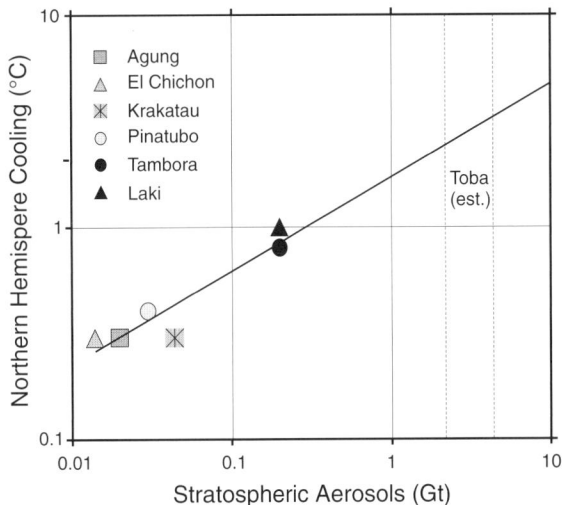

Figure 1. Temperature decreases in the Northern Hemisphere after volcanic eruptions plotted against the estimated loading of sulfuric acid aerosols in the stratosphere (data from Table 1). The estimated temperature decrease following the Toba eruption of ~73,500 yr B.P. is between 3 and 5 °C.

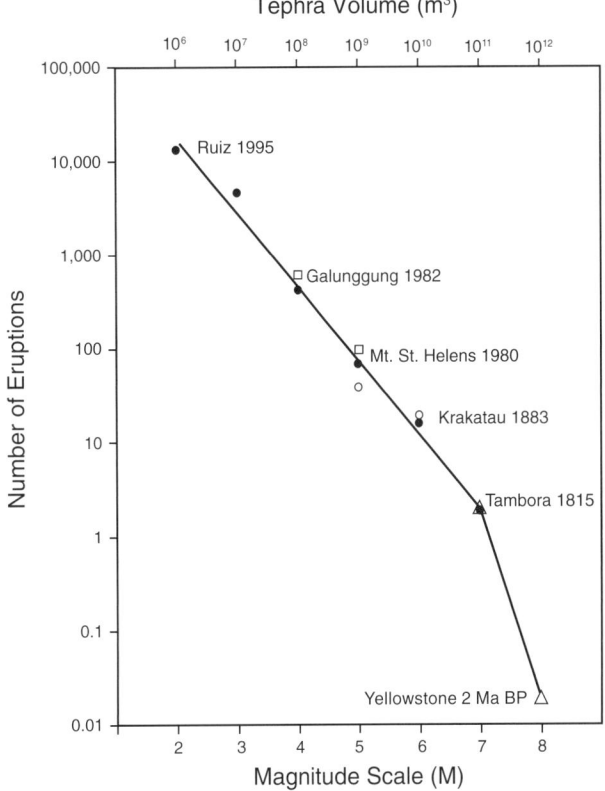

Figure 2. Plot of magnitude vs. frequency for volcanic eruptions (M classes 2–8). Total number of eruptions during various intervals (going back in time from 1 January 1994) for each M class is normalized to eruptions per 1000 years. Tephra volumes corresponding to each M class are shown at top. The best-fit line was determined by a regression model using the filled data points. Data points from Decker (1990) for M7 and M8 are shown by open triangles (modified after Simkin and Siebert, 1994).

35,000 years ago) contains the Campanian Ignimbrite (~70 km^3, M6) eruption (34,000 ± 3000 years ago), and another period of enhanced global volcanism around 80 Ka includes the Los Chocoyos tephra (>400 km^3, M7) from the Atitlan caldera (Rampino et al., 1979). The indications that volcanic eruptions cluster at times of global climatic change suggest that the >M8 supereruptions might also preferentially occur at these times.

Felsic Supereruptions and Humanity

With the real possibility of ongoing, human-induced, rapid climate change, it is worth considering the potential effect of a VEI 8 eruption on modern civilization. The primary effect would be through collapse of agriculture as a result of the loss of one or more growing seasons (Toon et al., 1997). As noted earlier, a Toba-scale (>M8) eruption is predicted to cause global cooling of 3–5 °C for several years, and regional cooling up to 15 °C (Rampino and Self, 1992, 1993; Rampino and Ambrose, 2000) (Fig. 1; Table 1). These deceptively mild values for cooling actually point to potential devastation of the major food-growing areas of the world. In Canada, a 2 to 3 °C average local temperature drop would destroy wheat production, and 3 to 4 °C would halt all Canadian grain production. Crops in the American Midwest and the Ukraine could be severely injured by a 3 to 4 °C temperature decrease (Harwell and Hutchinson, 1985; Pittock et al., 1986). Severe climate would also interfere with global transportation of foodstuffs and other goods. Thus, a Toba-sized eruption has the potential to compromise global agriculture, leading to famine and possible disease pandemics (Stothers, 1999).

The location of a supereruption can also be an important factor in its regional and global effects. Eruptions from the Yellowstone Caldera over the last 2 Ma include the 2500 km^3 Huckleberry Ridge Tuff (2 Ma), the 280 km^3 Mesa Falls Tuff (1.3 Ma), and the 1000 km^3 Lava Creek Tuff (0.6 Ma). Each of these produced thick ash deposits over the western and central United States (compacted ash thicknesses of 0.2 m occur ~1500 km from the source; Wood and Kienle, 1990; Sarna-Wojcicki and Davis, 1991). Thus, in addition to the global climatic consequences of these Yellowstone supereruptions, they would have directly devastated the major grain-producing areas in the breadbasket of North America, preventing agricultural recovery for years. As another example, if a Pacific eruption caused intense regional cooling, the Asian rice crop could be destroyed by even a single night of below-freezing temperatures during the growing season.

Furthermore, large volcanic eruptions might lead to longer-term climatic change through positive feedback effects on climate such as cooling the surface oceans, formation of sea ice, or increased land ice (Rampino and Self, 1992, 1993; Delworth et al., 2005), prolonging recovery from the "volcanic winter." The result of >M8 supereruptions could well be widespread starvation, famine, disease, social unrest, financial collapse, and severe damage to the underpinnings of civilization (Sagan and Turco, 1990).

Given these possible climatic and ecological effects, it is an interesting exercise to examine the potential for explosive supereruptions to limit our chances of finding intelligent extraterrestrial life. The chances for communicative intelligence in the Galaxy are commonly represented by a combination of the relevant factors called the Drake Equation, which can be written as

$$N = R^* f_p n_e f_l f_i f_c L, \qquad (1)$$

where N is the number of intelligent communicative civilizations in the Galaxy; R^* is the rate of star formation averaged over the lifetime of the Galaxy; f_p is the fraction of stars with planetary systems; n_e is the mean number of planets within such systems that are suitable for life; f_l is the fraction of such planets on which life actually occurs; f_i is the fraction of planets on which intelligence arises; f_c is the fraction of planets on which intelligent life develops a communicative phase; and L is the mean lifetime of such technological civilizations (Sagan, 1973).

Although the Drake Equation is useful in organizing the factors that are thought to be important for the occurrence of extraterrestrial intelligence, the actual assessment of the values of the terms in the equation is difficult. The only well-known number is R^*, which is commonly taken as 0.1 years. Consequently, estimates for N have varied from ~0 to >10^8 civilizations (Sagan, 1973).

It has been pointed out recently that f_c and L are limited in part by the occurrence of asteroid and comet impacts that could prove catastrophic to technological civilizations (Sagan and Ostro, 1994; Chyba, 1997). The threshold for global catastrophic climatic cooling and possible ozone layer destruction is estimated to be a 1- to 2-km-diameter asteroid or comet (Chapman and Morrison, 1994). Such an impact would release ~10^5 to 10^6 Mt (TNT equivalent) of energy, produce a crater ~20–40 km in diameter, and is calculated to generate a global cloud consisting of ~1 Gt of submicron dust (Toon et al., 1997). Impacts of this magnitude are expected to occur on average about every 10^5 years (Chapman and Morrison, 1994).

It is interesting that the effects of a >M8 eruption are strikingly similar to an ~1-km- diameter bolide impact, also injecting ~1 Gt of material into the atmosphere. Fine volcanic dust and sulfuric acid aerosols have optical properties similar to the submicron dust produced by impacts (Toon et al., 1997), and the effects' on atmospheric opacity should be similar. Volcanic aerosols, however, have a longer residence time of years (Bekki et al., 1996) compared with a few months for fine dust, so a >M8 eruption would be expected to have a longer-lasting effect on global climate than an impact producing a comparable amount of atmospheric loading.

Given that >M8 eruptions apparently occur more frequently than 1-km impacts (every 1–2 × 10^5 years versus 4–5 × 10^5 years) and potentially have even greater climate impact, volcanism could be the more important constraint on f_c and L in equation 1. Furthermore, the technology to detect and deflect threatening asteroids and comets is quite plausible. While the ability to forecast a >M8 supereruption is technologically feasible, a civilization would need to be far more advanced than ours to even contemplate preventing such a geologic event.

While preventing a volcanic eruption is impossible for us, we could take steps to enhance our ability to survive such a catastrophic event. One mitigation strategy could involve the stockpiling of global food reserves. In considering the vagaries of normal climatic change, Schneider and Londer (1984) noted that when grain stocks dip below ~15% of utilization, local scarcities, worldwide price jumps, and sporadic famine were more likely to occur. Indeed the current stockpile of grains is equivalent to a 2-month global supply (Smith, 2000), which is ~15% of annual consumption. For a volcanic VEI 8 catastrophe, however, several years of growing season might be curtailed (Zielinski et al., 1996a; Rampino and Ambrose, 2000), and hence a much larger stockpile of grain and other foodstuffs would have to be maintained, along with the means for efficient global distribution.

It does not take a detailed socioeconomic analysis to know that stockpiling 2 to 3 years of food is unrealistic under today's conditions. However, given sufficiently accurate forecasts several years in advance, it might be possible to mobilize the resources to avoid the posteruption collapse of civilization. Currently volcanic eruption forecasts are neither early enough nor accurate enough to allow the necessary mitigation efforts. Burrows and Shapiro (1999) suggest a different and more general solution for assuring the continuity of human civilization. They propose that space colonization transfer human civilization, along with all technological and cultural information, to other places in the Solar System for safekeeping. The repository would be a means of providing a backup system for the planet, fostering recovery of terrestrial civilization in the wake of global disasters of any type. This might sound like science fiction, but such a strategy may sadly be actually more feasible than having humanity respond appropriately before a >M8 supereruption.

MAFIC SUPERERUPTIONS ON EARTH

Mafic supereruptions (M8) are fundamentally different from felsic supereruptions. On Earth, they apparently only occur during discrete episodes of flood basalt volcanism, such as the Deccan Basalts in India (65 Ma) or the Columbia River Basalt Group (CRBG) in the northwestern United States (16 Ma). Such episodes are spaced tens of millions of years apart and generally last for only 0.5–1 Ma (e.g., Courtillot et al., 1986, 1999, 2000; Vandamme, et al., 1991; Coffin and Eldholm, 1994; Renne et al., 1995; Reichow et al., 2002; White and Saunders, 2005). As such, they are not a likely hazard to human civilization. However, as we will discuss later, flood volcanism is a common feature of all Earth-like planets and is therefore of much greater significance when one looks beyond the Earth and/or humanity.

Continental flood basalt provinces typically involve hundreds of individual flows, many with volumes ≥360 km^3 (M8). The total volumes of these flood basalt provinces exceed 10^6 km^3 implying that, on average, individual M8 eruptions occur about once every 10^4 years during the formation of the province (Rampino and Stothers, 1988). However, the formation of a province usually has a short waxing and peak period and a long waning period (e.g., Tolan et al., 1989; Courtillot et al., 1999). During the peak of the building of the province, M8 eruptions might be as frequent as every 10^3 years. In the following, we describe key components of flood basalt provinces and then discuss the atmospheric and climate effects of the associated eruptions.

Phreatomagmatic Volcaniclastic Deposits

There are two distinct types of events that dominate the construction of a flood basalt province: (1) phreatomagmatic and (2) flood lava eruptions. While the lava flows are more voluminous, the mafic volcaniclastic deposits (MVDs) formed via phreatomagmatic eruptions are also enormous. Recent work shows that MVDs are an important component of many flood basalt provinces including the Ferrar (e.g., Elliot and Hanson, 2001; White and McClintock, 2001), North Atlantic (e.g., Ukstins Peate et al., 2003; Larsen et al., 2003), Ontong Java (e.g., Thordarson, 2004), and Siberian traps (e.g., Ross et al., 2005). In most flood basalt provinces the MVDs are concentrated in the lower part of the succession and appear to be linked with the onset of province volcanism. These successions cover 10^3–10^5 km^2 and include tephra units with volumes of 10s–100s km^3 (M6-M7) (e.g., Ross et al., 2005). In several provinces (i.e., Karoo, Ferrar and Siberian traps), part of the MVD succession is characterized by thick units of tuff-breccias filling steep sided depressions up to 5 km in diameter that are interpreted to be very large diatremelike vent structures excavated by phreatomagmatic eruptions (e.g., White and McClintock, 2001). Elsewhere (i.e., North Atlantic, Ontong Java, and Ferrar), the succession is characterized by phreatomagmatic fall deposits that sometimes exhibit dispersal equivalent to that of phreatoplinian eruptions (e.g., Hanson and Elliot, 1996; Ukstins Peate et al., 2003; Jolly and Widdowson, 2005).

An example of these MVDs is the subaerial phreatomagmatic tuff and lapilli tuff succession recovered at ODP site 1184 on the Ontong Java Plateau (e.g., Thordarson, 2004). It consists of at least six 5–74-m-thick eruption units where stratified accretionary-lapilli tuff sequences alternating with diffuse-bedded lapilli tuffs (Fig. 3). Our current understanding of the eruption mechanics of these massive phreatomagmatic explosive eruptions is quite limited. However, the overall architecture of these units is that of a shower-bedded phreatomagmatic deposit analogous to that found in the proximal to medial fallout from the 1477 A.D. Veidivötn and 870 A.D. Vatnaöldur phreatomagmatic fissure eruptions in Iceland (Larsen, 1984). Continued research into these Icelandic eruptions may provide critical insight into this aspect of flood basalt eruptions. Of course, the issue of scaling up 1 to 2 orders of magnitude in volume will remain a challenge.

Flood Lava Flow Fields

Unlike the recently appreciated phreatomagmatic deposits, the massive lava flows that make up the bulk of each flood basalt province have been recognized essentially from the inception of geology. However, our understanding of how these M8 lava flow

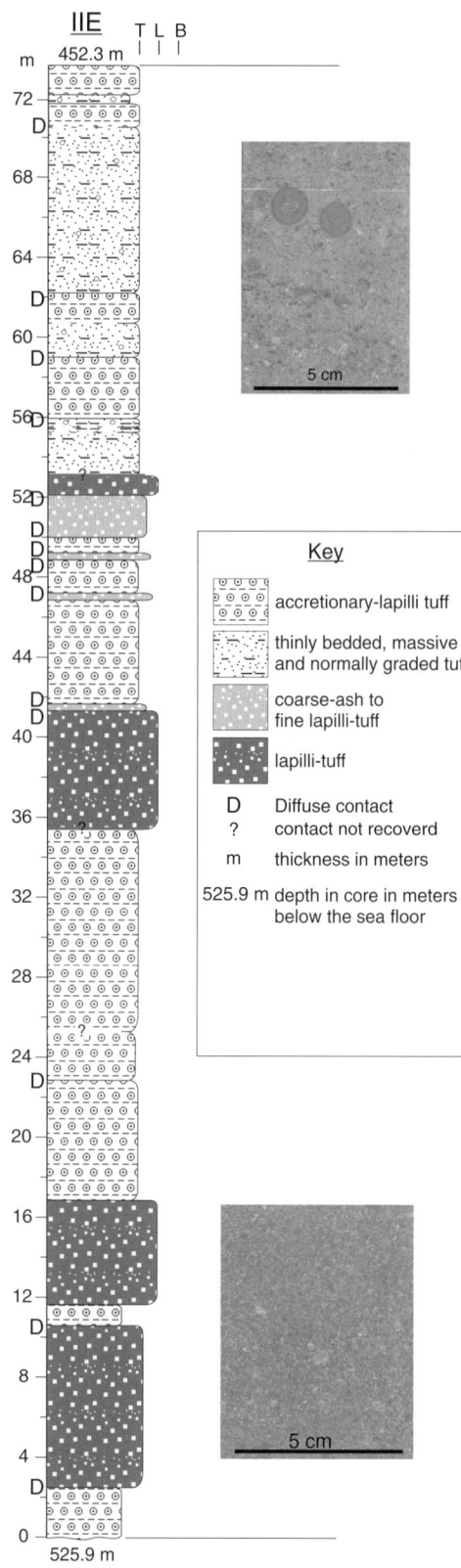

Figure 3. Stratigraphic log of phreatomagmatic eruption Unit IIE at ODP Site 1184 on the Ontong Java Plateau, showing typical lithofacies associations (after Thordarson, 2004).

fields were formed has changed dramatically over more than a century of study (e.g., Geikie, 1880; Washington, 1922; Tyrrell, 1937; Shaw and Swanson, 1970; Self et al., 1997). The most detailed work has come from the most recent flood basalt province, the ~16 Ma Columbia River Basalt Group. Work over the past 15 years shows that the products of individual eruptions are giant pahoehoe and rubbly pahoehoe flow fields produced by prolonged eruptions that most likely lasted for years to decades (Self et al., 1996, 1997, 1998; Keszthelyi and Self, 1998; Thordarson and Self, 1996, 1998; Keszthelyi et al., 2000, 2004, 2006). Each flow field consists of several lava flows, which in turn consist of multiple flow lobes. The major lobes are commonly 20–30 m thick, kilometers wide, and show classic inflation features. This same basic volcanic architecture has been reported from other flood basalt provinces where detailed physical descriptions have been published: Deccan Traps (e.g., Keszthelyi et al., 1999; Duraiswami et al., 2001, 2003; Bondre et al., 2000, 2004) and Etendeka (e.g., Jerram et al., 1999).

Both geothermometry and thermal modeling of the CRBG lava flows show that their length was not limited by cooling. The insulated transport of lava under a thick crust is extremely thermally efficient, with measured cooling rates of <0.1 °C per km traveled (Ho and Cashman, 1997; Thordarson and Self, 1998). This mode of emplacement has produced lava flow fields >1000 km long with an inferred lava flux of ~5000 m^3/s (Self et al., 2005). Thus the key to these "floods" of lava is the volume of the eruption, not the lava viscosity, eruption rate, or environmental conditions (Keszthelyi and Self, 1998).

Flood Basalt Vents

Important additional information about the nature of flood basalt eruptions can be gleaned from their vents. The frequent occurrence of dikes and dike systems that can be traced laterally for 10s–100s km imply that most flood basalt eruptions were fed by linear vent systems (Ernst and Buchan, 1997 and references therein). However, outcrops into vent constructs and near-vent succession are rare. The best documented examples are found within the CRBG (e.g., Swanson et al., 1975), where associations of fountain-fed flows, spatter and lapilli scoria units, and spongy and shelly pahoehoe lobes appear to define complex linear vent systems that bear a striking resemblance to modern-day fissure vents. One in-depth study of a vent within the ~16 Ma Teepee Butte Member of the Grande Ronde basalts shows that it featured a lava pond surrounded by ramparts that were constructed by at least three distinct episodes of Hawaiian-style fountaining (Reidel and Tolan, 1992).

Another detailed study of the near-vent succession of the ~14.7 Ma Roza Member at and around Winona, Washington, reveals a sequence where fountain-fed flows are overlain by 1–10-m-thick bedded lapilli scoria units, which in turn are capped by either fountain-fed lava or "normal" pahoehoe sheet lobes (Thordarson and Self, 1996). The beds exhibit features similar to the air fall deposits from the 1783–1784 A.D. Laki and

934–940 A.D. Eldgjá eruptions in Iceland (e.g., Thordarson et al., 2001). The Icelandic (and, presumably, Roza) deposits formed from distinct subplinian explosive phases at the beginning of individual eruption episodes. This stratigraphic sequence appears to be repeated several times in outcrops within the area, although more than one sequence is rarely present at any one site. Again, this pattern is identical to that found in the near-vent successions of the Laki and Eldgjá fissures, where individual sequences correspond to an eruption episode (Thordarson and Self, 1993; Thordarson et al., 2001, 2003a).

Although these studies are by no means conclusive regarding the mechanics of effusive flood basalt eruptions, the picture that is emerging conforms well to the notion that these fissure-fed lava-producing events are simply large-scale versions of the 15–20 km^3 historic eruptions in Iceland. Thus, by analogy, it is likely that they featured multiple eruption episodes, where each episode began with a relatively short-lived (days?) explosive phase followed by a longer-lasting effusive phase. Another important conclusion that can be drawn from this comparison is that it is unlikely that the entire vent system erupted concurrently. It is more likely that at any one time the activity was confined to distinct fissure segments on the vent system, as is evident from the mapping of the Roza lava flow field. This means that an eruption like Roza, which lasted for >10 years and had mean eruption rates of ~4000 m^3/s, would have been able to maintain 5–9-km-high columns throughout the eruption and could have supported up to 20-km-high columns during times of peak activity (Fig. 4).

Effect of Flood Basalt Eruptions on Climate

It has been suggested that flood basalt volcanism could be one of the extreme geologic events responsible for the "punctuation marks" in the history of life on Earth (e.g., Courtillot, 1999; Wignall, 2001; White and Saunders, 2005). While climate response to the release of magmatic gases has been assumed to provide the link to extinctions, this hypothesis is still not fully tested. Here, we present some initial steps toward that goal.

As with felsic supereruptions, there are several factors that control the magnitude of the climatic impact of large mafic eruptions. First, to produce a global effect the gas must be injected into the stratosphere; otherwise precipitation will remove the volcanic contribution quickly and effects will be only regional in extent. It is important to note that the height of the tropopause (the base of the stratosphere) is strongly latitude and climate dependent. Currently, it varies from 17 km at the equator to <10 km near the poles. As Figure 4 shows, M4–M8 mafic eruptions have the potential to penetrate the tropopause, but only during periods of peak magma discharge and under favorable atmospheric conditions.

Second, the longer climate is affected, the larger the impact on life. As shown for the felsic supereruptions, volcanic materials are cleansed from the stratosphere within a few years. This should also apply to the gases injected by mafic supereruptions. However, there is an enormous difference between the felsic and

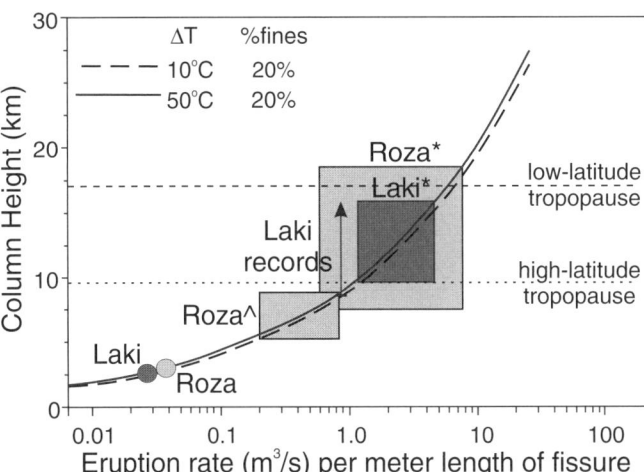

Figure 4. Height of eruption columns for M4–M8 flood lava eruptions using the fissure vent model of Stothers et al. (1986) as a function of eruption rate. The solid and dashed lines show the convective column height as a function of eruption rate for two values of temperature drop in fountaining tephra clasts (ΔT) and one value of fines content (ash < 1 mm) in the fountains. Circles labeled "Laki" and "Roza" denote the mean eruption rate divided by total length of fissure system. This greatly underestimates the actual column height because the entire fissure was never active at once. Box labeled "Roza^" marks the column heights calculated using the mean eruption rate divided by 5–20-km-long fissure segments. This is a reasonable estimate of the average column heights but does not take into account the fact that the peak eruption rate is typically 3–9 times greater than the mean eruption rate. Box labeled "Laki*" shows peak eruption rate divided by the actual length of erupting fissure. This matches the eyewitness records shown by the arrow. Box labeled "Roza*" shows the best estimate for the peak heights of the Roza eruption columns. The mean eruption rate is multiplied by a factor of 39 to estimate the peak eruption rate and fissure lengths of 5–20 km are assumed. At the mid-latitudes of the Roza eruption, the erupted gases are expected to enter the stratosphere only during the peak discharge of each episode of fissure eruptions.

mafic supereruptions: the felsic eruptions are relatively short-lived, with the bulk of the magma discharge completed within days; the mafic examples have repeated episodes every few months to years spread over several years to a few decades. Thus the full brunt of the climatic effect of a mafic supereruption is likely to persist unabated for many years or even decades.

Third, the single most significant factor controlling the magnitude of the climatic impact is the mass of H_2SO_4 aerosols injected into and produced in the stratosphere (and the upper troposphere). While ash and other volcanic gases also play an important role, the sulfuric acid aerosols are both the most persistent and the most climatically significant component (e.g., Thordarson and Self, 1996; Self et al., 2005, 2006). As the water released by these eruptions is 1.5–5 times the sulfur mass, it is reasonable to assume there is sufficient water available in the stratosphere and that the mass of H_2SO_4 aerosols is limited by the mass of sulfur released. While there are reasons to question this assumption (e.g., Thordarson and Self, 2003; Stevenson et al., 2003), it is a reasonable starting point. There are three methods for estimating sulfur release that we discuss briefly here: (1) simple extrapolation from smaller active eruptions, (2) the petrologic method, which compares the sulfur content of samples of the undegassed magma and degassed lava, and (3) geochemical proxies (especially Fe/Ti) for sulfur content in the magma (Thordarson et al., 1996, 2003a).

The best-studied mafic volcano is Kilauea on the island of Hawaii. Decades of monitoring have shown a reasonably consistent relationship between the volume of erupted lava and sulfur gas over the vent (Sutton et al., 2003). A value of 1.49 kg SO_2 per 1000 kg of magma provides an excellent fit to the data and is consistent with sulfur solubility in these melts. Simply scaling this relationship to a M8 eruption (1000 Gt) implies ~1.5 Gt of SO_2 being released into the atmosphere, which is more than in the Toba eruption.

The simple method has very significant limitations. The released gas may be confined to the troposphere and be incapable of global effects. In the case of Kilauea, the SO_2 aerosols generally do not travel more than a few hundred kilometers (e.g., Elias et al., 1998; McGee and Gerlach, 1998; Sutton et al., 2001), so the simple method might seriously overestimate the climatic impact. On the other hand, unlike large fissure eruptions, the magma feeding smaller eruptions often undergoes significant degassing in a shallow magma chamber. Thus, the simple method may underestimate the climate effects of a mafic supereruption.

A benchmark study that directly assessed the degassing of a mafic M8 eruption is that of Thordarson and Self (1996). The study included systematic measurements of sulfur in pristine melt inclusions within plagioclase phenocrysts and in quenched (glassy) tephra clasts as well as selvages from dikes and lava lobes of the Roza Member of the CRBG (Fig. 5). These samples record the progress of sulfur degassing during magma ascent and venting. Therefore, they provide estimates of total sulfur mass released into the atmosphere as well as the amount that could have been injected into the stratosphere. Approximately 90% of the sulfur carried in the magma was vented into the atmosphere by the eruption and >78% was released in fountains at the vents (Fig. 5). More than 9 Gt of SO_2 could have entered the stratosphere from this single eruption.

Note that this is about six times what is estimated by simply scaling up from Kilauea. Thordarson and Self (1996) demonstrated that the fact that flood lavas are fed by a subsurface plumbing system fundamentally different from that of shield volcano eruptions invalidates simple scaling from small historical eruptions.

Sulfur gas releases from flood basalt eruptions older than the ~16 Ma Columbia River Basalt Province are more difficult to

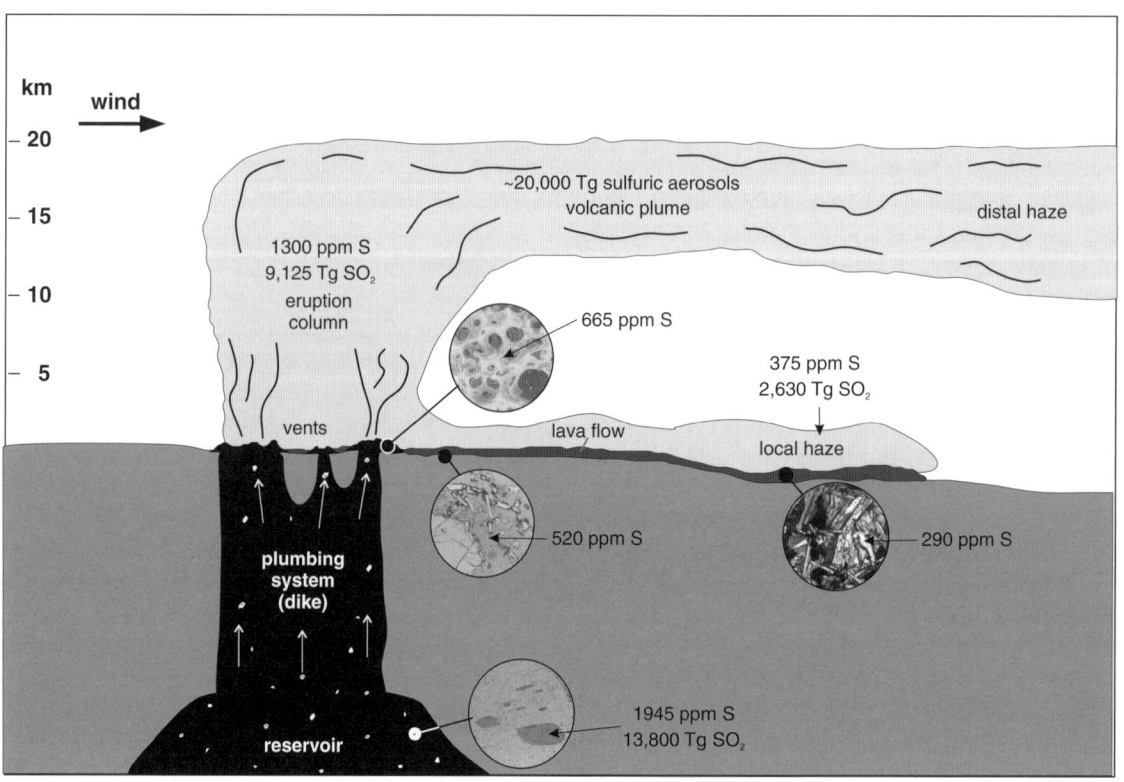

Figure 5. Measured degassing of the Roza Member of the CRBG. The measurements imply that the total amount of volatiles released by the Roza eruption was ~12 Gt SO_2, where >9 Gt SO_2 were liberated at the vents and an additional 2.6 Gt SO_2 were released by the lava during emplacement and cooling.

measure so directly because their glassy products have typically been weathered and altered. In instances where alteration renders application of the petrologic method unusable (for instance, the 65 Ma Deccan lavas of India), an empirical chemical proxy, using less mobile elements, can be used (Thordarson et al., 2003a; Self et al., 2005, 2006). Sulfur in lavas from the Eastern Volcanic Zone in Iceland exhibit a strong positive correlation with the TiO_2/FeO ratio (correlation coefficient $r = 0.83$). Sulfur in glass inclusions, magmatic tephra, and crystalline lava each define a distinct trend when plotted against the Ti-Fe ratio (Fig. 6). This relationship, combined with the fact that neither Fe nor Ti content is easily changed by moderate alteration (Rollinson, 1993), makes the TiO_2/FeO value an ideal proxy for estimating the original and the residual sulfur content of ancient flood lavas. Table 2 describes the best fit to each of these trends and to the amount released at each stage of the eruption. Thordarson et al. (2003a) showed that this empirical approach gives reliable estimates of the SO_2 budget, with an estimated error of ±15%.

The concentration of SO_2 in (or released from) the lava can be calculated by

$$MSO_2 = V_e \rho \varepsilon\, x_s, \quad (2)$$

where V_e is the volume of magma, ρ is the magma density, x_s is the mass fraction of sulfur, and ε is the ratio of the mass of S to SO_2 ($\varepsilon \approx 2$). Using this proxy method, Self et al. (2006) estimate that a typical 1000 km³ lava flow from the Deccan, Siberian, or Columbia River flood basalts had the potential to inject 5–10 Gt of SO_2 into the stratosphere. Since each of these lava flows probably took roughly a decade to be emplaced, this suggests that a M8 mafic eruption could be the equivalent of having a Toba-scale eruption every year for a decade. Even if only a fraction of the sulfur released at the vents actually reached the stratosphere, it is clear that flood basalt eruptions have the potential to cause atmospheric perturbations that could fundamentally change climate and even endanger life on Earth.

Flood Basalts and Mass Extinctions

It is tempting to assume that this "megascale" release of sulfur into the atmosphere provides the explanation for the puzzling coincidence in the timing of major extinction events and flood basalt eruptions. After all, the emplacement of a flood basalt province, with many hundreds of individual eruptions, would involve the injection of 1000–10,000 Gt of sulfur into the atmosphere, dwarfing the 40–600 Gt estimated for the Chicxulub impact at the K/T boundary (e.g., Yang and Ahrens, 1998; Courtillot and Thordarson, 2005). However, at the current time, a causative link between flood basalt eruptions and mass extinctions remains unclear. For example, there is no clear correlation between the magnitude of an extinction event and the volume of the temporally associated flood basalt province. This lack of recognizable correlation may be due to the simple fact that (a) we do not fully understand the factors that regulate the climate impact of these

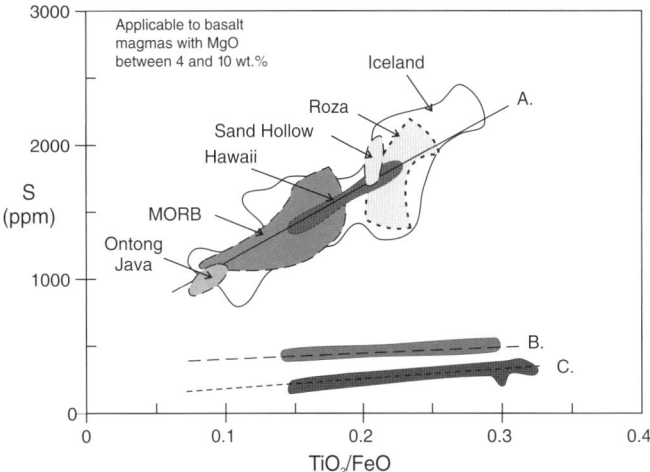

Figure 6. Sulfur vs. TiO_2/FeO ratio for various types of basalts. These data are used to create the empirical proxy method for calculating SO_2 release (Table 2). Trend A is the composition of melt inclusions that preserve the state of the unerupted magma. Trend B is the composition of the tephra, which preserves the state of the lava immediately after vent degassing. Trend C shows the composition of the solidified lava, after the flow had fully degassed. The difference between trends A and B is the loss of sulfur at the vent and the difference between trends B and C is the sulfur loss during emplacement and crystallization of lava flows. Modified from Thordarson et al. (2003a) and Self et al. (2006).

TABLE 2. EMPIRICAL MODEL FOR SULFUR RELEASE FROM MAFIC ERUPTIONS

Eruption stage	Sulfur concentration (ppm)
Magma (i.e., glass inclusions)	544 + 5456 TiO_2/FeO
Gas loss at the vent	185 + 4963 TiO_2/FeO
Lava fountains (i.e., tephra)	359 + 493 TiO_2/FeO
Gas loss from lava flows	177 + 14 TiO_2/FeO
Crystalline lava	183 + 479 TiO_2/FeO

eruptions and (b) we do not fully understand how the climate change(s) affected the biota of that time.

The magnitude of the sulfur release associated with a flood basalt eruption, and its approximately decade-long duration, is totally unprecedented in any existing models of volcanic gas release and sulfate aerosol generation (cf. Pinto et al., 1989). Large masses of volcanogenic SO_2 injected into the atmosphere will be converted to H_2SO_4 aerosols in the troposphere and stratosphere. In this way the SO_2 would have the potential to greatly deplete stratospheric H_2O and OH (Bekki, 1995; Bekki et al., 1996); the stratosphere currently only contains ~1 Gt of H_2O globally. How this enormous amount of aerosols would have manifested itself on atmospheric chemistry and dynamics cannot be currently calculated. The balance between a quasicontinuous volcanogenic source of S gas, the resulting H_2SO_4 aerosol formation rates, and deposition rates of the aerosols is only just beginning to be tested by atmospheric chemistry models (e.g., Stevenson et al., 2003; Chenet et al., 2005; Oman et al., 2006a). Preliminary work indicates that stratospheric dehydration would occur due to the high-altitude portion of the injected gas and delay the conversion to aerosols of further gas injection (Savarino et al., 2003). Estimates suggest that single flood basalt eruptions can release such large amounts of sulfate aerosols that they could create regional optical depths of >10. At such high levels, it is arguable that the immediate climate cooling might suppress convection in the lower atmosphere, to the extent that sulfur removal from the troposphere might become less effective than we might expect. Clearly, while the cooling effect resulting from a massive continuous flux of sulfur gas resulting from an average flood lava eruption can perhaps be crudely calculated (Jolly and Widdowson, 2005), complex global climate models will be required in order to properly explore the extent and severity of the effect.

While completely speculative at this time, it is irresistible to ponder how flood basalts may have played a key role in the evolution of life as we know it. Based on the admittedly crude estimates of the climatic impact of these eruptions, it appears that massive and sustained cooling of the biosphere is to be expected. Such climate change should drive some highly specialized species into extinction, but is unlikely to exterminate the majority of life. Instead, we speculate that flood basalt eruptions highly stress the ecosystem, killing large numbers of individuals across essentially all species. These remaining individuals would form relatively small and isolated ecosystems. If some other cataclysm occurs during this time of great stress, many of these small populations may be completely eliminated, resulting in a mass extinction event.

In the case of the K/T extinction event, there is no doubt that the Chicxulub impact event caused the single worst day for life on Earth in the past 100 Ma. The years that immediately followed the impact were also exceptionally harsh. However, had life been thriving at the time of the impact, it is not clear that small groups of many species might not have survived these hard times. Marine microbial life, in particular, should have been quite robust but was in no way spared in this extinction. However, if the eruption of Deccan flood lavas over the millennia prior to the impact had stressed life-forms to the degree that there were only small regions with isolated ecosystems, it would be much easier to eliminate viable populations of most species. Thus flood basalts may have loaded life on Earth with such hardship that the Chicxulub impact was the proverbial (and exceedingly large) straw that broke the camel's back.

MARTIAN MEGASCALE ERUPTIONS

Mars, the home of the largest volcanoes in the Solar System, has bountiful examples of large-scale eruptions. The famous shield volcanoes, some of which stand >20 km above the surrounding plateaus, are largely composed of lava flow fields with volumes of many tens of km^3 (e.g., Cattermole, 1990; Mouginis-Mark et al., 1992; Wilson and Head, 1994; Mouginis-Mark and Yoshioka, 1998). While large compared with common terrestrial events, their eruptions generally fall short of the threshold we use in this chapter for a "megascale" eruption (i.e., M8 or >360 km^3). Instead, as is the case for Earth, we will focus on flood lavas.

Global scale geomorphologic mapping suggests that much of the Martian plains are covered by flood lavas (e.g., Scott and Carr, 1978; Scott and Tanaka, 1986; Tanaka and Scott, 1987; Greeley and Guest, 1987; Tanaka et al., 2005). A very conservative estimate by Greeley and Schneid (1991) states that at least 38% of Mars's surface is covered by flood lavas. A more generous estimate, taking all the map units that have been described as likely to be flood lavas, suggests that more than half the current surface of Mars could be fresh or modified flood lava flow fields (Keszthelyi et al., 2006).

In some areas, the thickness of the lava stack appears to be several kilometers. McEwen et al. (1999) reported >8 km of layers in the outer slopes of Valles Marineris with the characteristic stair-step morphology of flood basalt provinces on Earth. Schultz (2002) showed that the canyon wall-rock strengths are consistent with those of layered igneous rocks. However, Beyer and McEwen (2005) suggested that much of the overall wall strength in Coprates Chasma may be controlled by a few especially strong layers, so the majority of the layers could be much weaker, perhaps rich in tephra. The layers in Valles Marineris that are interpreted to be flood lavas are distinct from the finer layers seen inside the canyons and elsewhere (Malin and Edgett, 2001). Ori and Karna (2003) have mapped out the global distribution of thick-bedded units; they are concentrated around Tharsis (including Valles Marineris) and Elysium, the major volcanic regions of Mars.

Perhaps most intriguing is the indication that flood volcanism has continued into the most recent period of Mars's geologic history. A region in the Elysium and Amazonis plains, ~3000 × 700 km, is covered with relatively fresh looking flood lavas (Plescia, 1990). A review of the uncertainties in the ages determined from the size-frequency distribution of craters on these young surfaces allows for the youngest lavas to have erupted 1.5–200 Ma ago (McEwen et al., 2005). The whole province formed during the last ~700 Ma (Plescia, 1990). While locally cut by water floods and covered by various mantling deposits, the surface of these youngest flood lavas are generally well-exposed and essentially pristine in 1–10 m/pixel images.

In many ways, the best data set with which to view these lava flows is the global digital elevation model (DEM) provided by the Mars Orbital Laser Altimeter (MOLA). The data were originally acquired in strips under the orbiting spacecraft, with 170-m-diameter laser spots spaced 300 m apart. The data were then used to produce a 128 pixel/degree (~500 m/pixel at the equator) DEM of the planet, with better than 1 m vertical precision (Smith et al., 2001). Individual lava flows, ~10 m thick, can readily be seen extending for hundreds of kilometers across the Elysium plains (Plescia, 2003; Sakimoto and Gregg, 2004; Keszthelyi et al., 2004) (Fig. 7). They are fed from the Cerberus Fossae, a system of parallel fissures extending for more than 1000 km. A group of

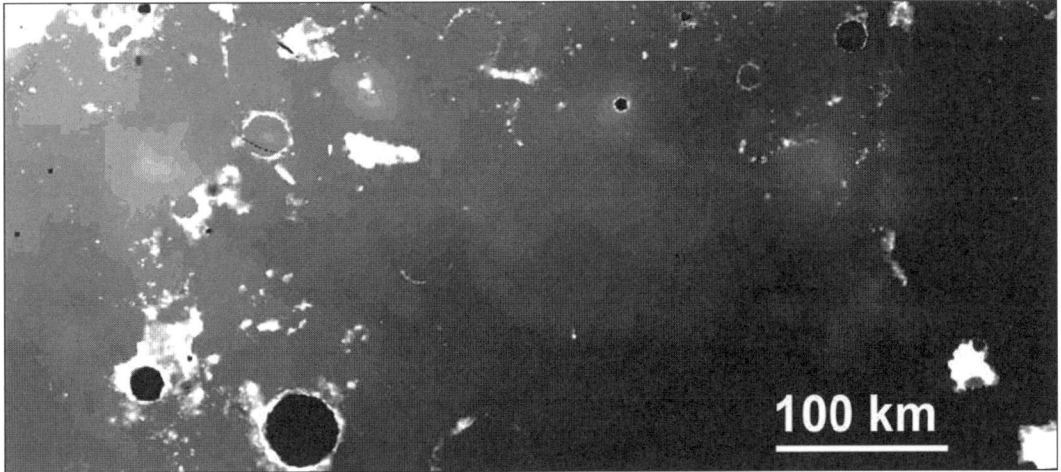

Figure 7. Mars Orbiter Laser Altimeter (MOLA) digital elevation model of flood lavas in Elysium Planitia. Mercator projection over 3–9 °N, 158–172 °E. Elevations are shown in a linear grayscale between –2500 and –3000 m. Note the 100–300-km-long lava flows with lobate margins in the center of the figure. About a dozen low shields are also visible, concentrated along the Cerberus Fossae fissure system. Larger flows, some of which appear to extend for >1500 km, can also be discerned in other parts of the MOLA data.

low shields (provisionally named the Cerberus Tholi) sit astride the fissures, marking locations where the effusion of lava became concentrated at point sources during extended eruptions. While it is not possible to link every lava flow to a specific vent, the most distant flow termini are ~1500 km from the main group of low shields. With widths approaching 100 km, the volume of some of these flows must be close to 1000 km^3, firmly indicating that M8 flood lava flows do exist on Mars.

Platy-Ridged Lava and Rubbly Pahoehoe

The slow rate of geologic processes on Mars provides a unique view of the upper surfaces of megascale lava flow fields, even though they are tens to hundreds of millions of years old. These largest lava flow fields on Mars do not feature surface structures typical of inflated pahoehoe lavas. Instead the Martian flood lava surfaces are characterized by rafted plates and pressure ridges, initially leading to the nongenetic name "platy-ridged terrain" (Keszthelyi et al., 2000; Malin and Edgett, 2001). While some researchers suggest that platy-ridged terrain is formed by freezing mudflows (Rice et al., 2002; Williams and Malin, 2004) or freezing water (Murray et al., 2005), the evidence that these are lava flows is overwhelming (Keszthelyi et al., 2004; Lanagan, 2004).

The Martian flood lavas exhibit a set of morphological features that are identical in all aspects, except scale, to some lava flow fields in Iceland (Keszthelyi et al., 2004; Guilbaud et al., 2005) (Fig. 8). These features include not only rafted plates and compressional ridges, but also parallel grooves carved as wakes into the flow top as it translated over topographic obstacles. The cross-cutting relationships between the ridges and plates require that some areas underwent repeated breakups. Also, inflated pahoehoe are found along the margins of individual flows, both in Iceland and on Mars. The paucity of water in the upper meter of platy-ridged terrain (as measured by the Gamma Ray Spectrometer onboard the Mars Global Surveyor spacecraft [Boynton et al., 2002]) and the lack of sublimation-related degradation of the surface in high-resolution observations effectively rules out formation mechanisms involving a water-rich fluid.

Simple models for the emplacement of these flows, and the observed sizes, suggest that eruption rates were on the order of 10^4 m^3/s, with surges in the range of 10^5 m^3/s or more (Keszthelyi et al., 2004). Even with such high eruption rates, it would have taken years to emplace any M8 lava flow. However, new studies of the youngest flood lava flow field on Mars show that it drapes channel-like (flood erosion?) features, and evidence suggests that it may have been emplaced in a turbulent fashion over a period of only a few weeks (Jaeger et al., 2007, 2008). The possibility that lava flow fields on Mars may have been produced by eruptions lasting, on one hand, for years to decades and on the other hand, for several weeks has important implications for assessment of eruption styles and thus their potential atmospheric impact.

Evidence for Explosive Eruptions

While spectroscopic studies have suggested the presence of some intermediate composition lavas on Mars (e.g., McSween et al., 1999; Christensen et al., 2001), there has been no evidence for felsic volcanism on the planet. Therefore any explosive volcanism on Mars is assumed to be mafic. Theoretical studies suggest that eruptions on Mars should be more explosive than on Earth, primarily due to the thin atmosphere on Mars. This low ambient pressure (<1% of Earth's) allows gas to expand further, leading to a larger gas volume fraction (and

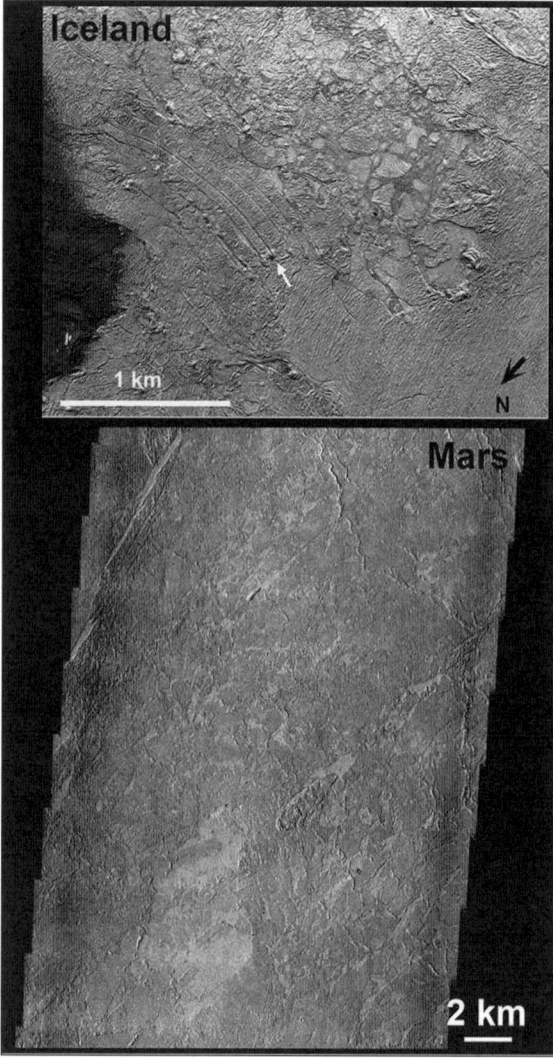

Figure 8. Comparison of Icelandic and Martian platy-ridged lava surfaces. The first image is an aerial photo from the 1783–1784 Laki flow field in Iceland. Note the rafted plates, compressional ridges (flow top rubble pushed up into ridges oriented perpendicular to the direction of flow), and wakes (claw-mark like structures above the white arrow). The arrow points to the start of a wake and indicates the predominant direction of crust motion. The second image is a portion of THEMIS Vis image number V10242005 centered at 5.5 °N, 175 °E in Elysium Planitia. Note the set of morphologic features similar to those seen on Laki, albeit at a larger scale.

thus degree of fragmentation) in the same magma ascending on Mars as compared to the Earth (Wilson and Head, 1994). However, observational evidence of explosive eruptions on Mars has been somewhat indirect. In part, this is because the predicted pyroclastic deposits are essentially indistinguishable from the dust that covers most of the planet. It is unclear what observations would allow a feature to be unambiguously identified as an explosive volcanic vent.

Despite this challenge, there are many likely examples of deposits from explosive eruptions on Mars. These include large sections of older major volcanic constructs including Hecates Tholus, Tyrrhena Patera, and Hadriaca Patera (e.g., Mouginis-Mark et al., 1992). However, it is difficult to ascertain whether or not individual eruptions at these volcanoes involved >1000 Gt of lava and would be considered "megascale" for this chapter.

The most likely example of an ash deposit large enough to require eruptions of that scale is the Medusae Fossae Formation (MFF). The MFF is a relatively young unit characterized by a layered deposit that is easily eroded by the wind. It is found in large patches across a region of ~1000 × 6000 km, extending along the equator west of the Tharsis shield volcanoes (e.g., Scott and Tanaka, 1982). Other similar deposits are found on the other side of the planet (Hynek et al., 2003), but they are probably significantly older (e.g., Scott and Tanaka, 1986; Malin and Edgett, 2001). While many hypotheses have been put forward for the origin of the MFF, volcanic ash has been generally favored (e.g., Tanaka et al., 2003). The source of the ash has been assumed to be the giant Tharsis shield volcanoes, especially since the MFF is thicker closer to these volcanoes. Modeling the atmospheric transport of ash lofted by an eruption plume suggests that such transport is plausible, but only if the ash is micron scale (Hynek et al., 2003). Keszthelyi et al. (2000) suggest a closer source could be the young flood lavas in Elysium Planitia. The model of Hynek et al. (2003) would allow 10-μm particles to reach this far from the Cerberus Fossae source region. Preferential deposition and reduced erosion rates at the higher elevations on the flanks of the volcanoes might explain the observed distribution of the MFF. It is also likely that much of the deposit has been significantly modified (eroded and redeposited) by aeolian processes.

Estimating the size of the eruptions that might have built up the MFF has many uncertainties. The MFF currently covers ~2 × 10^6 km^2, but has suffered significant erosion. At the same time, redeposition could have increased the aerial extent of the deposit. While it is unlikely that all the eruptions would have covered the entire current extent of the MFF, it is plausible that some did. Layering within the MFF can be found on many scales, from <5 m to >500 m. If we conservatively assume that a 1–10-m-thick layer represents the products of a single eruption, we estimate 2 ×10^3 – 2 × 10^4 km^3 of ash per eruption (M8 to M9). This admittedly crude analysis suggests that the MFF was produced by hundreds, or even thousands, of eruptions. Given the age constraints on the MFF, these eruptions could have been distributed across nearly 3 Ga. However, if they are associated with the younger flood lavas to the north, the MFF may have formed in the last 700 Ma. If the association with the flood lavas is correct, it also suggests that the volume of pyroclastics may have been roughly 1–10 times the volume of the effusive lavas. While terrestrial fissure eruptions tend to have roughly ten times more lava than pyroclastics (e.g., Thordarson and Self, 1993), a far larger volume of pyroclastics is expected on Mars (e.g., Wilson and Head, 1994; Wilson and Mouginis-Mark, 2003).

Improved Volatile Release Estimates

It has been postulated that these massive eruptions could have had dramatic effects on the Martian atmosphere and climate. Plescia (1993) estimated that the eruption of the group of flood lavas that covered the Elysium plains could have produced ~10^{16} kg of water and a similar amount of CO_2. This is enough CO_2 to double the atmospheric pressure and enough water to blanket all of Mars in a layer of ice ~10 cm deep. However, as noted in the previous section, this kind of input was probably spread over hundreds or thousands of eruptions across almost a billion years. What might be the consequence of a single eruption?

The largest flood lava eruptions on Mars may have involved as much as 10^4 km^3 of magma. Table 3 shows the estimated SO_2 release calculated using the empirical method for terrestrial flood lavas and the Fe/Ti ratios from various samples of Martian lava. We estimate as much as 20–40 Gt of SO_2 could be expected to be released over one of these multiyear eruptions. Mafic magmas typically contain 4 to 5 times as much water as SO_2, so 100–240 Gt of highly acidic aerosols would be expected from the largest Martian eruptions. While this is roughly 1% of the mass of the atmosphere, it only amounts to about 1 mm of precipitation if spread globally. It is clear that fluvial erosion features seen on Mars cannot be the direct consequence of even the largest volcanic eruptions. Instead, it is likely that the ultimate fate of the erupted gases is to be precipitated as a sulfuric acid–rich frost on the surface. The sublimation of the water would leave behind a sulfate deposit. Given that we expect these large eruptions to have persisted for more than a Mars year, and since the winds on Mars shift dramatically over the course of a year, these deposits should be distributed essentially globally. Indeed, volcanic aerosols are hypothesized to be a key ingredient in forming the sulfates seen in the widespread dust deposits across Mars (e.g., Settle, 1979; Bullock and Moore, 2004).

The effect of these volcanic gasses while they are in the atmosphere is currently not fully understood. However, it is likely that the sulfur and water would combine into acidic aerosols. While the atmospheric composition is quite different, the temperatures and pressures in Mars's lower atmosphere are remarkably similar to those in the Earth's stratosphere, so it is plausible that aerosol formation would be similar on both planets. However, micron-scale particles would fall out of the Martian atmosphere on a time scale of only months, as opposed to years for Earth. So, if an M8 or M9 eruption on Mars releases 100–240 Gt of water and SO_2 over a period of years, we would expect an order 1 Gt of aerosols (i.e., equivalent to a typical dust storm on Mars, see below) to be suspended in the atmosphere at any given time. However, in case of short-lived, high-discharge (turbulent) flood lava eruptions, the instantaneous atmospheric loading would be increased to perhaps 100 to 1000 times what is typical of major dust storms on Mars (e.g., Jaeger et al., 2008).

Overall, the climate effect of these aerosols would be broadly similar to that on the Earth, capturing solar energy within the atmosphere and cooling the surface. Mars's climate undergoes similar changes during the global dust storms that occur some years. In fact, the dust loading in the atmosphere during these events is estimated to be 0.3–0.6 Gt (e.g., Cantor et al., 2001), comparable to the ~1 Gt expected from volcanic aerosols. Thus, because Mars naturally undergoes massive interannual climate change, it is unclear that even the largest volcanic eruptions would cause atmospheric conditions significantly outside the normal variability.

TABLE 3. ESTIMATED SULFUR RELEASE FROM MARTIAN LAVAS

Sample	TiO/FeO	Gt SO_2/1000 km^3
Nakhla	0.017	2.1
Chassignites	0.003	1.8
QUE 94201	0.099	4.1
Pathfinder "sulfur-free rock"	0.058	3.1
MER (Gusev Crater lava)	0.034	2.5

Note: Compositions from Lodders (1998); McSween et al. (1999, 2004).

CONCLUSIONS

Megascale (>M8) volcanic eruptions, involving >1000 Gt of magma, are some of the most powerful geologic processes that affect both Earth and Mars. On Earth, felsic megaeruptions recur on average about every 100–200,000 years and have the potential to be catastrophic to human civilization. The 73.5 ka Toba eruption may have had an indelible impact on human evolution. Any future eruption of this type will require an unprecedented level of coordinated response from humankind to assure our continuation as a space-faring species. Mafic M8 eruptions (i.e., flood basalts events) are focused in periods of time sufficiently infrequent to be of no real concern to humans. However, the massive injection of volcanic gases (especially SO_2) may have played a part in punctuating the evolution of life on this planet. The interplay between the more sustained effects of flood basalt eruptions and the (geologically) instantaneous effects of large impacts is worthy of more detailed study than has been completed to date.

On Mars, mafic eruptions larger than any in Earth's geologic record are well preserved. While the effects of these eruptions were undoubtedly global, the climate impact may not have been outside what Mars endures even in the absence of volcanic eruptions. This is testament to the extreme variability of the current Martian atmosphere, rather than a lack of power in these eruptions.

ACKNOWLEDGMENTS

This study was supported in part by NSF grant ATM-0313965. We extend our thanks to Mary Chapman for her extended encouragement and support during the course of this work, and Lionel Wilson for his constructive review and helpful suggestions.

REFERENCES CITED

Ambrose, S.H., 1998, Late Pleistocene human population bottlenecks, volcanic winter, and differentiation of modern humans: Journal of Human Evolution, v. 34, p. 623–651, doi: 10.1006/jhev.1998.0219.

Bekki, S., 1995, Oxidation of volcanic SO_2: A sink for stratospheric OH and H_2O: Geophysical Research Letters, v. 22, p. 913–916, doi: 10.1029/95GL00534.

Bekki, S., Pyle, J.A., Zhong, W., Toumi, R., Haigh, J.D., and Pyle, D.M., 1996, The role of microphysical and chemical processes in prolonging the climate forcing of the Toba eruption: Geophysical Research Letters, v. 23, p. 2669–2672, doi: 10.1029/96GL02088.

Beyer, R.A., and McEwen, A.S., 2005, Layering stratigraphy of eastern Coprates and northern Capri Chasmata, Mars: Icarus, v. 179, p. 1–23, doi: 10.1016/j.icarus.2005.06.014.

Bondre, N.R., Dole, G., Phadnis, V.M., Duraiswami, R., and Kale, V.S., 2000, Inflated pahoehoe lavas from the Sangamner area of the western Deccan Volcanic Province: Current Science, v. 78, p. 1004–1007.

Bondre, N.R., Duraiswami, R.A., and Dole, G., 2004, Morphology and emplacement of flows from the Deccan Volcanic Province, India: Bulletin of Volcanology, v. 66, p. 29–45, doi: 10.1007/s00445-003-0294-x.

Boynton, W.V., Feldman, W.C., Squyres, S.W., Prettyman, T.H., Brückner, J., Evans, L.G., Reedy, R.C., Starr, R., Arnold, J.R., Drake, D.M., Englert, P.A.J., Metzger, A.E., Mitrofanov, I., Trombka, J.I., d'Uston, C., Wänke, H., Gasnault, O., Hamara, D.K., Janes, D.M., Marcialis, R.L., Maurice, S., Mikheeva, I., Taylor, G.J., Tokar, R., and Shinohara, C., 2002, Distribution of Hydrogen in the near surface of Mars: Evidence for subsurface ice deposits: Science, v. 297, p. 81–85, doi: 10.1126/science.1073722.

Bullock, M.A., and Moore, J.M., 2004, Aqueous alteration of Mars-analog rocks under acid atmosphere: Geophysical Research Letters, v. 31, doi: 10.1029/2004GL019980.

Burrows, W.E., and Shapiro, R., 1999, An alliance to rescue civilization: A bold proposal for Earth's future: Ad Astra., Sept./Oct., p. 18–22.

Cantor, B.A., James, P.B., Caplinger, M., and Wolff, M.J., 2001, Martian dust storms: 1999 Mars Orbiter Camera observations: Journal of Geophysical Research, v. 106, p. 23,653–23,688, doi: 10.1029/2000JE001310.

Carrecedo, J.C., Rodriguez-Badiloa, E., and Soler, V., 1992, The 1730–1736 eruption of Lanzarote, Canary Islands: A long, high-magnitude basaltic fissure eruption: Journal of Volcanology and Geothermal Research, v. 53, p. 239–250, doi: 10.1016/0377-0273(92)90084-Q.

Cattermole, P., 1990, Volcanic flow development at Alba Patera, Mars: Icarus, v. 83, p. 453–493, doi: 10.1016/0019-1035(90)90079-O.

Chapman, C.R., and Morrison, D., 1994, Impacts on the Earth by asteroids and comets: Assessing the hazards: Nature, v. 367, p. 33–40, doi: 10.1038/367033a0.

Chenet, A.L., Fluteau, F., and Courtillot, V., 2005, Modelling massive sulphate aerosol pollution, following the large 1783 Laki basaltic eruption: Earth and Planetary Science Letters, v. 236, p. 721–731, doi: 10.1016/j.epsl.2005.04.046.

Chesner, C.A., Rose, W., Deino, A., Drake, R., and Westgate, J.A., 1991, Eruptive history of the Earth's largest Quaternary caldera (Toba, Indonesia) clarified: Geology, v. 19, p. 200–203, doi: 10.1130/0091-7613(1991)019<0200:EHOESL>2.3.CO;2.

Christensen, P.R., Bandfield, J.L., Hamilton, V.E., Ruff, S.W., Kieffer, H.H., Titus, T.N., Malin, M.C., Morris, R.V., Lane, M.D., Clark, R.L., Jakosky, B.M., Mellon, M.T., Pearl, J.C., Conrath, B.J., Smith, M.D., Clancy, R.T., Kuzmin, R.O., Roush, T., Mehall, G.L., Gorelick, N., Bender, K., Murray, K., Dason, S., Greene, E., Silverman, S., and Greenfield, M., 2001, Global Surveyor Thermal Emission Spectrometer experiment: Investigation description and surface science results: Journal of Geophysical Research, v. 106, p. 23,823–23,972, doi: 10.1029/2000JE001370.

Chyba, C.F., 1997, Catastrophic impacts and the Drake Equation, in Cosmovici, C.B., Bowyer, S., and Werthimer, D., eds., Astronomical and Biochemical Origins and the Search for Life in the Universe: Bologna, Editrice Compositori, p. 157–164.

Coffin, M.F., and Eldholm, O., 1994, Large igneous provinces: Crustal structure, dimensions, and external consequences: Reviews of Geophysics, v. 32, p. 1–36, doi: 10.1029/93RG02508.

Courtillot, V., 1999, Evolutionary Catastrophes: Cambridge, UK, Cambridge University Press, 188 p.

Courtillot, V., and Thordarson, T., 2005, Flood basalts appear to be the main cause of biological mass extinctions in the Phanerozoic: Geophysical Research Abstracts, European Geoscience Union General Assembly, Vienna, Abstract# EGU05-A-11196.

Courtillot, V., Besse, J., Vandamme, D., Montigny, R., Jaegger, J.J., and Capetta, H., 1986, Deccan flood basalts at the Cretaceous/Tertiary boundary?: Earth and Planetary Science Letters, v. 80, p. 361–374, doi: 10.1016/0012-821X(86)90118-4.

Courtillot, V., Jaupart, C., Manighetti, I., Tapponnier, P., and Besse, J., 1999, On causal links between flood basalts and continental breakup: Earth and Planetary Science Letters, v. 166, p. 177–195, doi: 10.1016/S0012-821X(98)00282-9.

Courtillot, V., Gallet, Y., Rocchia, R., Feraud, G., Robin, E., Hofmann, C., Bhandari, N., and Ghevariya, Z.G., 2000, Cosmic markers, Ar-40/Ar-39 dating and paleomagnetism of the KT sections in the Anjar Area of the Deccan large igneous province: Earth and Planetary Science Letters, v. 182, p. 137–156, doi: 10.1016/S0012-821X(00)00238-7.

Decker, R.W., 1990, How often does a Minoan eruption occur? in Hardy, DA, ed., Thera and the Aegean World III: London, UK, Thera Foundation, v. 2, p. 444–452.

De La Cruz-Reyna, S., 1991, Poisson-distributed patterns of explosive eruptive activity: Bulletin of Volcanology, v. 54, p. 57–67, doi: 10.1007/BF00278206.

Delworth, T.L., Ramaswamy, V., and Stenchikov, G.L., 2005, The impact of aerosols on simulated ocean temperature and heat content in the 20th century: Geophysical Research Letters, v. 32, p. L24709, doi: 10.1029/2005GL024457.

Demaree, G.R., Ogilvie, A.E.J., and Zhang, D., 1998, Further documentary evidence of Northern Hemispheric coverage of the Great Dry Fog of 1783: Climatic Change, v. 39, p. 727–730, doi: 10.1023/A:1005319607233.

Duraiswami, R.A., Bondre, N.R., Dole, G., Phadnis, V.M., and Kale, V.S., 2001, Tumuli and associated features from the western Deccan Volcanic Province, India: Bulletin of Volcanology, v. 63, p. 435–442, doi: 10.1007/s004450100160.

Duraiswami, R.A., Dole, G., and Bondre, N.R., 2003, Slabby pahoehoe from the western Deccan Volcanic Province: Evidence for incipient pahoehoe-aa transitions: Journal of Volcanology and Geothermal Research, v. 121, p. 195–217, doi: 10.1016/S0377-0273(02)00411-0.

Elias, T., Sutton, A.J., Stokes, J.B., and Casadevall, T.J., 1998, Sulfur dioxide emission rates of Kilauea Volcano, Hawaii, 1979–1997: U.S. Geological Survey Open-File Report 98–462, 41 p.

Elliot, D.H., and Hanson, R.E., 2001, Origin of widespread, exceptionally thick basaltic phreatomagmatic tuff breccia in the middle Jurassic Prebble and Mawson Formations, Antarctica: Journal of Volcanology and Geothermal Research, v. 111, p. 183–201, doi: 10.1016/S0377-0273(01)00226-8.

Ernst, R.E., and Buchan, K.L., 1997, Giant radiating dyke swarms: Their use in identifying Pre-Mesozoic large igneous provinces and mantle plumes, in Mahoney, J.J., and Coffin, M.F., eds., Large Igneous Provinces: Continental, Oceanic, and Planetary Flood Volcanism, American Geophysical Union Geophysical Monograph, Washington, D.C., v. 100, p. 297–334.

Fiacco, R.J., Jr., Thordarson, T., Germani, M., Self, S., Palais, J.M., Withlow, S., and Grootes, P.M., 1994, Atmospheric aerosol loading and transport due to the 1783–1784 Laki eruption in Iceland, interpreted from ash particles and acidity in the GISP2 ice core: Quaternary Research, v. 42, p. 231–240, doi: 10.1006/qres.1994.1074.

Geikie, A., 1880, The lava-fields of north-western Europe: Nature, v. 23, p. 3–5, doi: 10.1038/023003a0.

Glazner, A.F., Manley, C.R., Marron, J.S., and Rojstaczer, S., 1999, Fire or ice: Anticorrelation of volcanism and glaciation in California over the past 800,000 years: Geophysical Research Letters, v. 26, p. 1759–1762, doi: 10.1029/1999GL900333.

Godano, C., and Civetta, L., 1996, Multifractal analysis of Vesuvius volcano eruptions: Geophysical Research Letters, v. 23, p. 1167–1170, doi: 10.1029/96GL00966.

Grattan, J., and Brayshay, M., 1995, An amazing and portentous summer: Environmental and social responses in Britain to the 1783 eruption of an Iceland volcano: The Geographical Journal, v. 161, p. 125–134, doi: 10.2307/3059970.

Greeley, R., and Guest, J.E., 1987, Geologic Map of the Eastern Equatorial Region of Mars, U.S. Geological Survey Miscellaneous Investigations Map I-1802-B, 1:15,000,000 scale.

Greeley, R., and Schneid, B.D., 1991, Magma generation on Mars: Amounts, rates, and comparisons with Earth, Moon, and Venus: Science, v. 254, p. 996–998, doi: 10.1126/science.254.5034.996.

Guilbaud, M.-N., Self, S., Thordarson, T., and Blake, S., 2005, Morphology, surface structures, and emplacement of lavas produced by Laki, A.D. 1783–1784, in Manga, M., and Ventura, G., eds., Kinematics and dynamics of lava flows: Geological Society of America Special Paper 396, p. 81–102, doi: 10.1130/2005.2396(07).

Hálfdánarson, G., 1984, Mannfall í kjölfar Skaftárelda (Loss of human lives following the Laki eruption), in Einarsson et al., eds., Skaftáreldar 1783–84: Ritgerdir og Heimildir, Mál og Menning, Reykjavík, p. 139–162.

Hanson, R.E., and Elliot, D., 1996, Rift-Related Jurassic Basaltic Phreatomagmatic Volcanism in the Central Trans-Antarctic Mountains: Precursory Stage to Flood-Basalt Effusion: Bulletin of Volcanology, v. 58, p. 327–347, doi: 10.1007/s004450050143.

Hardarson, B.S., and Fitton, J.G., 1991, Increased mantle melting beneath Snaefellsjökull volcano during Late Pleistocene glaciation: Nature, v. 353, no. 6339, p. 62–64, doi: 10.1038/353062a0.

Harington, C.R., ed., 1992, The Year Without a Summer? World Climate in 1816, Canadian Museum of Nature, Ottawa, Canada, 576 p.

Harpending, H.C., Sherry, S.T., Rodgers, A.L., and Stoneking, M., 1993, The genetic structure of ancient human populations: Current Anthropology, v. 34, p. 483–496, doi: 10.1086/204195.

Harwell, M.A., and Hutchinson, T.C., eds., 1985, Environmental Consequences of Nuclear War, v. II: Ecological and Agricultural Effects: New York, Wiley, 523 p.

Ho, C.H., 1991, Non-homogeneous Poisson model for volcanic eruptions: Mathematical Geology, v. 23, p. 167–173, doi: 10.1007/BF02066293.

Ho, A.M., and Cashman, K.V., 1997, Temperature constraints on the Ginkgo Flow of the Columbia River Basalt Group: Geology, v. 25, p. 403–406, doi: 10.1130/0091-7613(1997)025<0403:TCOTGF>2.3.CO;2.

Hynek, B.M., Phillips, R.J., and Arvidson, R.E., 2003, Explosive volcanism in the Tharsis region: Global evidence in the Martian geologic record: Journal of Geophysical Research, v. 108, doi: 10.1029/2003JE002062.

Jaeger, W.L., Keszthelyi, L.P., McEwen, A.S., Titus, T.N., Dundas, C.M., and Russell, P.S., 2007, Athabasca Valles, Mars: A lava-draped channel system: Science, v. 317, p. 1709–1711.

Jaeger, W.L., Keszthelyi, L., McEwen, A.S., and the HiRISE Team, 2008, Emplacement of Athabasca Valles flood lavas: Lunar and Planetary Science Conference, 39, Abstract #1836.

Jerram, D., Mountney, N., Holzforster, F., and Stollhofen, H., 1999, Internal stratigraphic relationships in the Etendeka Group in the Huab Basin, NW Namibia: Understanding the onset of flood volcanism: Journal of Geodynamics, v. 28, p. 393–418, doi: 10.1016/S0264-3707(99)00018-6.

Jolly, W.J., and Widdowson, M., 2005, Did Paleogene North Atlantic rift-related eruptions drive early Eocene climate cooling?: Lithos, v. 79, p. 355–366, doi: 10.1016/j.lithos.2004.09.007.

Keszthelyi, L., and Self, S., 1998, Some physical requirements for the emplacement of long basaltic lava flows: Journal of Geophysical Research, v. 103, p. 27,447–27,464, doi: 10.1029/98JB00606.

Keszthelyi, L., Self, S., and Thordarson, T., 1999, Applications of recent studies on the emplacement of basaltic lava flows to the Deccan Traps: Memoir of the Geological Society of India, v. 43, p. 485–520.

Keszthelyi, L., McEwen, A.S., and Thordarson, T., 2000, Terrestrial analogs and thermal models for Martian flood lavas: Journal of Geophysical Research, v. 105, p. 15,027–15,049, doi: 10.1029/1999JE001191.

Keszthelyi, L., Thordarson, T., McEwen, A.S., Haack, H., Guilbaud, M.N., Self, S., and Rossi, M., 2004, Icelandic analogs to Martian flood lavas: Geochemistry, Geophysics, Geosystems, v. 5, 2004GC000758.

Keszthelyi, L., Self, S., and Thordarson, T., 2006, Flood lavas on Earth, Io, and Mars: Journal of the Geological Society, v. 163, p. 253–264, doi: 10.1144/0016-764904-503.

Lanagan, P.D., 2004, Geologic history of the Cerberus Plains, Mars [Ph.D. thesis]: Tucson, University of Arizona, 142 p.

Larsen, G., 1984, Recent volcanic history of the Veidivotn fissure swarm, southern Iceland - An approach to volcanic risk assessment: Journal of Volcanology and Geothermal Research, v. 22, p. 33–58, doi: 10.1016/0377-0273(84)90034-9.

Larsen, G., 2000, Holocene eruptions within the Katla volcanic system, south Iceland: Characteristics and environmental impact: Jökull, v. 49, p. 1–28.

Larsen, L.M., Fitton, J.G., and Pedersen, A.K., 2003, Paleogene volcanic ash layers in the Danish Basin: Compositions and source areas in the North Atlantic igneous province: Lithos, v. 71, p. 47–80, doi: 10.1016/j.lithos.2003.07.001.

Lodders, K., 1998, A survey of shergottite, nakhlite and chassigny meteorites whole-rock compositions: Meteoritics & Planetary Science, v. 33, p. A183–A190.

Malin, M.C., and Edgett, K.S., 2001, Mars Global Surveyor Mars Orbiter Camera: Interplanetary cruise through primary mission: Journal of Geophysical Research, v. 106, p. 23,429–23,570, doi: 10.1029/2000JE001455.

Maclennan, J., Jull, M., McKenzie, D., Slater, L., and Gronvold, K., 2002, The link between volcanism and deglaciation in Iceland: Geochemistry Geophysics Geosystems, v. 3, no. 11, p. 1062, doi: 10.1029/2001GC000282.

McEwen, A.S., Malin, M.C., Carr, M.H., and Hartmann, W.K., 1999, Voluminous volcanism on early Mars revealed in Valles Marineris: Nature, v. 397, p. 584–586, doi: 10.1038/17539.

McEwen, A.S., Preblich, B.S., Turtle, E.P., Atremieva, N.A., Golombek, M.P., Hurst, M., Kirk, R.L., Burr, D.M., and Christensen, P.R., 2005, The rayed crater Zunil and interpretations of small impact craters on Mars: Icarus, v. 176, p. 351–381, doi: 10.1016/j.icarus.2005.02.009.

McGee, K.A., and Gerlach, T.M., 1998, Airborne volcanic plume measurements using a FTIR spectrometer, Kilauea volcano, Hawaii: Geophysical Research Letters, v. 25, p. 615–618, doi: 10.1029/98GL00356.

McGuire, W.J., Howarth, R.J., Firth, C.R., Solow, A.R., Pullen, A.D., Saunders, S.J., Stewart, I.S., and Vita-Finzi, C., 1997, Correlation between rate of sea-level change and frequency of explosive volcanism in the Mediterranean: Nature, v. 389, p. 473–476, doi: 10.1038/38998.

McSween, H.Y., Murchie, S.L., Crisp, J.A., Bridges, N.T., Anderson, R.C., Bell, J.F., III, Britt, D.T., Brueckner, J., Dreibus, G., Economou, T., Ghosh, A., Golombek, M.P., Greenwood, J.P., Johnson, J.R., Moore, H.J., Morris, R.V., Parker, T.J., Rieder, R., Singer, R.B., and Waenke, H., 1999, Chemical, multispectral, and textural constraints on the composition and origin of rocks at the Mars Pathfinder landing site: Journal of Geophysical Research, v. 104, p. 8679–8715, doi: 10.1029/98JE02551.

McSween, H.Y., Arvidson, R.E., Bell, J.F., III, Blaney, D., Cabrol, N.A., Christensen, P.R., Clark, B.C., Crisp, J.A., Crumpler, L.S., Des Marais, D.J., Farmer, J.D., Gellert, R., Ghosh, A., Gorevan, S., Graff, T., Grant, J., Haskin, L.A., Herkenhoff, K.E., Johnson, J.R., Jolliff, B.L., Klingelhoefer, G., Knudson, A.T., McLennan, S., Milam, K.A., Moersch, J.E., Morris, R.V., Rieder, R., Ruff, S.W., de-Souza, P.A., Squyres, S.W., Waenke, H., Wang, A., Wyatt, M.B., Yen, A., and Zipfel, J., 2004, Basaltic rocks analyzed by the Spirit rover in Gusev Crater: Science, v. 305, p. 842–845, doi: 10.1126/science.3050842.

Mouginis-Mark, P.J., and Yoshioka, M., 1998, The long lava flows of Elysium Planitia, Mars: Journal of Geophysical Research, v. 103, p. 19,389–19,400, doi: 10.1029/98JE01126.

Mouginis-Mark, P.J., Wilson, L., and Zuber, M.T., 1992, The physical volcanology of Mars, in Kieffer, H.H., Jakosky, B.M., Snyder, C.W., and Matthews, M.S., eds., Mars: Tucson, Univ. Arizona Press, p. 424–452.

Murray, J.B., Mueller, J.-P., Neukum, G., Werner, S.C., van Gassalt, S., Hauberk, E., Markiewicz, W.J., Head, J.W., III, Foing, B.H., Page, D., Mitchell, K.L., Portyankina, G., Albertz, J., Basilevsky, A.T., Bellucci, G., Bibring, J.P., Buchroithner, M., Carr, M.H., Dorrer, E., Duxbury, T.C., Ebner, H., Greeley, R., Helpke, C., Hoffmann, H., Inada, A., Ip, W.H., Ivanov, B.A., Jaumann, R., Keller, H.U., Kirk, R., Kraus, K., Kronberg, P., Kuzmin, R., Langevin, Y., Lumme, K., Masson, P., Mayer, H., McCord, T.B., Neubauer, F.M., Oberst, J., Ori, G.G., Paetzold, M., Pinet, P., Pischel, R., Poulet, F., Raitala, J., Schwarz, G., Spohn, T., and Squyres, S.W., 2005, Evidence from the Mars Express High Resolution Stereo Camera for a frozen sea close to Mars' equator: Nature, v. 434, p. 352–356, doi: 10.1038/nature03379.

Newhall, C.A., and Self, S., 1982, The volcanic explosivity index (VEI): An estimate of the explosive magnitude for historical volcanism: Journal of Geophysical Research, v. 87, p. 1231–1238, doi: 10.1029/JC087iC02p01231.

Oman, L., Robock, A., Stenchikov, G.L., Thordarson, T., Koch, D., Shindell, D.T., and Gao, C., 2006a, Modeling the Distribution of the Volcanic Aerosol Cloud from the 1783–1784 Laki Eruption: Journal of Geophysical Research, v. 111, no. D12, doi: 10.1029/2005JD006899.

Oman, L., Robock, A., Stenchikov, G.L., and Thordarson, T., 2006b, High-latitude eruptions cast shadow over the African monsoon and the flow of the Nile: Geophysical Research Letters, v. 33, doi:10.1029/2006GL027665.

Ori, G.G., and Karna, A., 2003, The uppermost crust of Mars and flood basalts: Lunar and Planetary Science Conference XXXIV, abstract 1539.

Paterne, M., and Guichard, F., 1993, Triggering of volcanic pulses in the Campanian area, South Italy, by periodic deep magma influx: Journal of Geophysical Research, v. 98, p. 1861–1873, doi: 10.1029/92JB02662.

Paterne, M., Labeyrie, J., Guichard, F., Mazaud, A., and Maitre, F., 1990, Fluctuations of the Campanian explosive volcanic activity (South Italy) during the past 190,000 years, as determined by marine tephrachronology: Earth and Planetary Science Letters, v. 98, p. 166–174, doi: 10.1016/0012-821X(90)90057-5.

Pinto, J.P., Turco, R.P., and Toon, O.B., 1989, Self-limiting physical and chemical effects in volcanic eruption clouds: Journal of Geophysical Research, v. 94, p. 11,165–11,174, doi: 10.1029/JD094iD08p11165.

Pittock, A., Ackerman, T., Crutzen, P., McCracken, M., Shapiro, C., and Turco, R., eds., 1986, Environmental Consequences of Nuclear War, v. I: Physical and Atmospheric Effects, New York, Wiley, 359 p.

Plescia, J.B., 1990, Recent flood lavas in the Elysium region of Mars: Icarus, v. 88, p. 465–490, doi: 10.1016/0019-1035(90)90095-Q.

Plescia, J.B., 1993, An assessment of volatile release from recent volcanism in Elysium Mars: Icarus, v. 104, p. 20–32, doi: 10.1006/icar.1993.1079.

Plescia, J.B., 2003, Cerberus Fossae, Elysium, Mars: A source for lava and water: Icarus, v. 164, p. 79–95, doi: 10.1016/S0019-1035(03)00139-8.

Post, J.D., 1978, The last great subsistence crisis in the western world: Baltimore, Maryland, Johns Hopkins University Press, 69 p.

Pyle, D.M., 1995, Mass and energy budgets of explosive volcanic eruptions: Geophysical Research Letters, v. 22, p. 563–566, doi: 10.1029/95GL00052.

Pyle, D.M., Beattie, P.D., and Bluth, G.J.S., 1996, Sulphur emissions to the stratosphere from explosive volcanic eruptions: Bulletin of Volcanology, v. 57, p. 663–671, doi: 10.1007/s004450050119.

Rampino, M.R., and Ambrose, S., 2000, Volcanic winter in the Garden of Eden: The Toba super-eruption and the Late Pleistocene human population crash: Geological Society of America Special Paper, v. 345, p. 71–82.

Rampino, M.R., and Self, S., 1984, Sulphur-rich volcanism and stratospheric aerosols: Nature, v. 310, p. 677–679, doi: 10.1038/310677a0.

Rampino, M.R., and Self, S., 1992, Volcanic winter and accelerated glaciation following the Toba super-eruption: Nature, v. 359, p. 50–52, doi: 10.1038/359050a0.

Rampino, M.R., and Self, S., 1993, Climate-volcanism feedback and the Toba eruption of ~74,000 years ago: Quaternary Research, v. 40, p. 269–280, doi: 10.1006/qres.1993.1081.

Rampino, M.R., Self, S., and Fairbridge, R.W., 1979, Can rapid climatic change cause volcanic eruptions?: Science, v. 206, p. 826–829, doi: 10.1126/science.206.4420.826.

Rampino, M.R., and Stothers, R.B., 1988, Flood basalt volcanism during the past 250 million years: Science, v. 241, p. 663–668, doi: 10.1126/science.241.4866.663.

Rampino, M.R., Stothers, R.B., and Self, S., 1988, Volcanic winters: Annual Review of Earth and Planetary Sciences, v. 16, p. 73–99, doi: 10.1146/annurev.ea.16.050188.000445.

Reichow, M., Saunders, A.D., White, R.V., Pringle, M.A., Al'Mukhamedov, A., Medvedev, A., and Kirda, N.P., 2002, $^{40}Ar/^{39}Ar$ dates from the West Siberian Basin: Siberian Flood Basalt Province doubled: Science, v. 296, p. 1846–1849, doi: 10.1126/science.1071671.

Reidel, S.P., and Tolan, T.L., 1992, Eruption and emplacement of flood basalt: An example from the large-volume Teepee Butte Member: Columbia River Basalt Group: Geological Society of America Bulletin, v. 104, p. 1650–1671.

Renne, P.R., Zhang, Z., Richards, M.A., Black, M.T., and Basu, A.R., 1995, Synchrony and causal relations between Permian–Triassic boundary crises and Siberian flood volcanism: Science, v. 269, p. 1413–1416, doi: 10.1126/science.269.5229.1413.

Rice, J.W., Jr., Parker, T.J., Russel, A.J., and Knudsen, O., 2002, Morphology of fresh outflow channel deposits on Mars: 33rd Lunar and Planetary Science Conference, CD-ROM XXXIII, abstract no. 2026.

Rollinson, H., 1993, Using geochemical data: Evaluation, presentation, interpretation: Essex, England, Longman Scientific & Technical, 351 p.

Ross, P.-S., Ukstins Peate, I., McClintock, M.K., Xu, Y.G., Skilling, I.P., White, J.D.L., and Houghton, B.F., 2005, Mafic volcaniclastic deposits in flood basalt provinces: A review: Journal of Volcanology and Geothermal Research, v. 145, p. 281–314, doi: 10.1016/j.jvolgeores.2005.02.003.

Sagan, C., ed., 1973, Communication with Extraterrestrial Intelligence: Cambridge, Massachusetts, MIT Press, 428 p.

Sagan, C., and Ostro, S., 1994, Long-range consequences of interplanetary collisions: Issues in Science and Technology, v. 10, p. 67–72.

Sagan, C., and Turco, R., 1990, A Path Where No Man Thought: New York, New York, Random House, 499 p.

Sakimoto, S.E.H., and Gregg, T.K.P., 2004, Cerberus Fossae and Elysium Planitia lavas, Mars: Source vents, flow rates, edifice styles, and water interactions: 35th Lunar and Planetary Science Conference, CD-XXXV, abstract no. 1851.

Sarna-Wojcicki, A.M., and Davis, J.O., 1991, Quaternary Tephrochronology: The Geology of North America, v. K2, p. 93–116.

Savarino, J., Bekki, S., Cole-Dai, J., and Thiemens, M.H., 2003, Evidence from sulfate mass independent oxygen isotopic compositions of dramatic changes in atmospheric oxidation following massive volcanic eruptions: Journal of Geophysical Research, v. 108, doi: 10.1029/2003JD003737.

Schneider, S.H., and Londer, R., 1984, The Coevolution of Climate and Life: San Francisco, California, Sierra Club, 563 p.

Scott, D.H., and Carr, M.H., 1978, Geologic Maps of Mars: U.S. Geological Society Miscellaneous Investigations Map I-1803.

Scott, D.H., and Tanaka, K.L., 1982, Ignimbrites of Amazonis Planitia region of Mars: Journal of Geophysical Research, v. 87, p. 1179–1190, doi: 10.1029/JB087iB02p01179.

Scott, D.H., and Tanaka, K.L., 1986, Geologic Map of the Western Equatorial Region of Mars: U.S. Geological Survey Miscellaneous Investigations Map I-1802-A, 1:15,000,000 scale.

Schultz, R.A., 2002, Stability of rock slopes in Valles Marineris, Mars: Geophysical Research Letters, v. 29, p. 38–41.

Self, S., 2006, The effects and consequences of very large explosive volcanic eruptions: Phil. Trans. Royal Soc. A, v. 364, p. 2073–2097; doi:10.1098/rsta.2006.1814.

Self, S., Thordarson, T., Keszthelyi, L., Walker, G.P.L., Hon, K., Murphy, M.T., Long, P., and Finnemore, S., 1996, A new model for the emplacement of the Columbia River Basalt as large, inflated pahoehoe sheet lava flow fields: Geophysical Research Letters, v. 23, p. 2689–2692, doi: 10.1029/96GL02450.

Self, S., Thordarson, T., and Keszthelyi, L., 1997, Emplacement of continental flood basalt lava flows, in Mahoney, J.J., and Coffin, M.F., eds., Large Igneous Provinces: Continental, Oceanic, and Planetary Flood Volcanism: Washington, D.C., American Geophysical Union, v. 100, p. 381–410.

Self, S., Keszthelyi, L., and Thordarson, T., 1998, The importance of pahoehoe: Annual Review of Earth and Planetary Sciences, v. 26, p. 81–110, doi: 10.1146/annurev.earth.26.1.81.

Self, S., Gertisser, R., Thordarson, T., Rampino, M.R., and Wolff, J.A., 2004, Magma volume, volatile emissions, and stratospheric aerosols from the 1815 eruption of Tambora—revisited: Geophysical Research Letters, v. 31, doi:10.1029/2004GL020925.

Self, S., Thordarson, T., and Widdowson, M., 2005, Flood basalt eruptions: Gas fluxes and potential effects on the global environment: Elements, v. 1, p. 283–288, doi: 10.2113/gselements.1.5.283.

Self, S., Widdowson, M., Thordarson, T., and Jay, A.E., 2006, Volatile fluxes during flood basalt eruptions and potential effects on the global environment: A Deccan perspective: Earth and Planetary Science Letters, v. 248, p. 517–531.

Settle, M., 1979, Formation and deposition of volcanic sulfate aerosols on Mars: Journal of Geophysical Research, v. 84, p. 8348–8354.

Shaw, H.R., and Swanson, D.A., 1970, Eruption and flow rates of flood basalts: Proceedings of the Second Columbia River Basalt Symposium: Cheney, Washington, Eastern Washington State College Press, p. 271–299.

Sigvaldason, G.E., Annertz, K., and Nilsson, M., 1992, Effect of glacier loading/unloading on volcanism: Postglacial volcanic production rate of the Dyngjufjöll area, central Iceland: Bulletin of Volcanology, v. 54, p. 385–392, doi: 10.1007/BF00312320.

Simkin, T., and Siebert, L., 1994, Volcanoes of the World: Tucson, Arizona, Geoscience Press, 349 p.

Smith, D.E., Zuber, M.T., Frey, H.V., Garvin, J.B., Head, J.W., Muhleman, D.O., Pettengill, G.H., Phillips, R.J., Solomon, S.C., Zwally, H.J., Banerdt, W.B., Duxbury, T.C., Golombek, M.P., Lemoine, F.G., Neumann, G.A., Rowlands, D.D., Aharonson, O., Ford, P.G., Ivanov, A.B., Johnson, C.L., McGovern, P.J., Abshire, J.B., Afzal, R.S., and Sun, X., 2001, Orbiter Laser Altimeter: Experiment summary after the first year of global mapping of Mars: Journal of Geophysical Research, v. 106, p. 23,689–23,722, doi: 10.1029/2000JE001364.

Smith, Z.A., 2000, The Environmental Policy Paradox: Upper Saddle River, New Jersey, Prentice Hall, 283 p.

Sparks, S., Self, S., Pyle, D., Oppenheimer, C., Rymer, H., and Grattan, J., 2005, Super-eruptions: Global effects and future threats: Report of a Geological Society of London Working Group, www.geolsoc.org.uk/supereruptions.

Stevenson, D.S., Johnson, C.E., Highwood, E.J., Gauci, V.Y., Collins, W.J., and Derwent, R.G., 2003, Atmospheric impact of the 1783–1784 Laki eruption: Part I Chemistry modeling: Atmospheres: Chemical Physics, v. 3, p. 487–507.

Stothers, R.B., 1983, The great Tambora eruption of 1815 and its aftermath: Science, v. 224, p. 1191–1198.

Stothers, R.B., 1984, Mystery cloud of AD 536: Nature, v. 307, p. 344–345, doi: 10.1038/307344a0.

Stothers, R.B., 1996, The great dry fog of 1783: Climatic Change, v. 32, p. 79–89, doi: 10.1007/BF00141279.

Stothers, R.B., 1998, Far reach of the Tenth Century Eldgja eruption, Iceland: Climatic Change, v. 39, p. 715–726, doi: 10.1023/A:1005323724072.

Stothers, R.B., 1999, Volcanic dry fogs, climate cooling, and plague pandemics in Europe and the Middle East: Climatic Change, v. 42, p. 713–723, doi: 10.1023/A:1005480105370.

Stothers, R.B., 2000, Climatic and demographic consequences of the massive volcanic eruption of 1258: Climatic Change, v. 45, p. 361–374, doi: 10.1023/A:1005523330643.

Stothers, R.B., Wolff, J.A., Self, S., and Rampino, M.R., 1986, Basaltic fissure eruptions, plume heights, and atmospheric aerosols: Geophysical Research Letters, v. 13, p. 725–728, doi: 10.1029/GL013i008p00725.

Sutton, A.J., Elias, T., Gerlach, T.M., and Stokes, J.B., 2001, Implications for eruptive processes as indicated by sulfur dioxide emission from Kilauea volcano, Hawaii, USA, 1979–1997: Journal of Volcanology and Geothermal Research, v. 108, p. 283–302, doi: 10.1016/S0377-0273(00)00291-2.

Sutton, A.J., Elias, T., and Kauahikaua, J., 2003, Lava-effusion rates for the Puu Oo—Kuapaianaha eruption derived from SO2 emissions and Very Low Frequency (VLF) measurements: U.S. Geological Survey Professional Paper, v. 1676, p. 137–148.

Swanson, D.A., Wright, T.L., and Helz, R.T., 1975, Linear vent systems and estimated rates of magma production and eruption for the Yamika basalt on the Columbia Plateau: American Journal of Science, v. 275, p. 877–905.

Tanaka, K.L., and Scott, D.H., 1987, Geologic Map of the Polar Regions of Mars: U.S. Geological Survey Miscellaneous Investigations Map I1802-C, 1:15,000,000 scale.

Tanaka, K.L., Skinner, J.A., Hare, T.M., Joyal, T., and Wenker, A., 2003, Resurfacing history of the northern plains of Mars based on geologic mapping of Mars Global Surveyor data: Journal of Geophysical Research, v. 109, doi: 10.1029/2002JE001908.

Tanaka, K.L., Skinner, J.A., and Hare, T.M., 2005, Geologic Map of the Northern Plains of Mars: U.S. Geological Survey Miscellaneous Investigations Map I-2888, 1:5,000,000 scale.

Thordarson, T., 2004, Accretionary-lapilli-bearing pyroclastic rocks at ODP Leg 192 Site 1184: A record of subaerial phreatomagmatic eruptions on the Ontong Java Plateau, in Fitton, J. G., Mahoney, J. J., Wallace, P. J., and Saunders, A. D., eds., Origin and Evolution of the Ontong Java Plateau: Special Publication of the Geological Society, v. 229, p. 275–306.

Thordarson, T., and Self, S., 1993, The Laki (Skaftar Fires) and Grimsvotn eruptions in 1783–1785: Bulletin of Volcanology, v. 55, p. 233–263, doi: 10.1007/BF00624353.

Thordarson, T., and Self, S., 1996, Sulfur, chlorine and fluorine degassing and atmospheric loading by the Roza eruption, Columbia River Basalt Group, Washington, USA: Journal of Volcanology and Geothermal Research, v. 74, p. 49–73, doi: 10.1016/S0377-0273(96)00054-6.

Thordarson, T., and Self, S., 1998, The Roza Member, Columbia River Basalt Group: A gigantic pahoehoe lava flow field formed by endogenous processes?: Journal of Geophysical Research, v. 103, no. B11, p. 27,411–27,445, doi: 10.1029/98JB01355.

Thordarson, T., and Self, S., 2003, Atmospheric and environmental effects of the 1783–84 Laki eruption: Journal of Geophysical Research, v. 108, no. D1, p. 4011, doi: 10.1029/2001JD002042.

Thordarson, T., Self, S., Óskarsson, N., and Hulsebosch, T., 1996, Sulfur, chlorine and fluorine degassing and atmospheric loading by the 1783–1784 AD Laki (Skaftár Fires) eruption in Iceland: Bulletin of Volcanology, v. 58, p. 205–225, doi: 10.1007/s004450050136.

Thordarson, T., Miller, D.J., Larsen, G., Self, S., and Sigurdsson, H., 2001, New estimates of sulfur degassing and atmospheric mass-loading by the 934 AD Eldgjá eruption, Iceland: Journal of Volcanology and Geothermal Research, v. 108, p. 33–54, doi: 10.1016/S0377-0273(00)00277-8.

Thordarson, T., Self, S., Miller, D.J., Larsen, G., and Vilmundardóttir, E.G., 2003a, Sulphur release from flood lava eruptions in the Veidivötn, Grímsvötn and Katla volcanic systems, Iceland, in Oppenheimer, C., Pyle, D. M., and Barclay, J., eds., Volcanic Degassing: Special Publication of the Geological Society, v. 213, p. 103–121.

Thordarson, T., Larsen, G., Steinthorsson, S., and Self, S., 2003b, 1783–85AD Laki-Grímsvötn eruptions II: Appraisal based on contemporary accounts: Jökull, v. 51, p. 11–48.

Tolan, T.L., Reidel, S.P., Beeson, M.H., Anderson, J.L., Fecht, K.R., and Swanson, D.A., 1989, Revision to the estimates of the aerial extent and volume of the Columbia River Basalt Group, in Reidel, S. D., and Hooper, P. R., eds., Volcanism and Tectonism in the Columbia River flood-basalt province: Geological Society of America Special Paper 239, p. 1–20.

Toon, O.B., Turco, R.P., and Covey, C., 1997, Environmental perturbations caused by the impacts of asteroids and comets: Reviews of Geophysics, v. 35, p. 41–78, doi: 10.1029/96RG03038.

Tyrrell, G.W., 1937, Flood basalts and fissure eruption: Bulletin of Volcanology, v. 1, p. 87–111.

Ukstins Peate, I., Larsen, M., and Lesher, C.E., 2003, The transition from sedimentation to flood volcanism in the Kangerlussuaq Basin, East Greenland: Basaltic pyroclastic volcanism during initial Palaeogene continental break-up: Journal of the Geological Society, v. 160, p. 759–772, doi: 10.1144/0016-764902-071.

Vandamme, D., Courtillot, V., Montigny, R., and Besse, J., 1991, Paleomagnetism and age determinations of the Deccan traps (India): Results of the Nagpur–Bombay traverse and review of earlier work: Reviews of Geophysics, v. 29, p. 159–190, doi: 10.1029/91RG00218.

Washington, H.S., 1922, Deccan Traps and the other plateau basalts: Geological Society of America Bulletin, v. 33, p. 765–804.

White, J.D.L., and McClintock, M.K., 2001, Immense vent complex marks flood-basalt eruption in a wet, failed rift: Coombs Hills, Antarctica: Geology, v. 29, p. 935–938, doi: 10.1130/0091-7613(2001)029<0935:IVCMFB>2.0.CO;2.

White, R.V., and Saunders, A.D., 2005, Volcanism, impact and mass extinctions: Incredible or credible coincidences?: Lithos, v. 79, p. 299–316, doi: 10.1016/j.lithos.2004.09.016.

Wignall, P.B., 2001, Large igneous provinces and mass extinctions: Earth-Science Reviews, v. 53, p. 1–33, doi: 10.1016/S0012-8252(00)00037-4.

Williams, R.M.E., and Malin, M.C., 2004, Evidence for late stage fluvial activity in Kasei Valles, Mars: Journal of Geophysical Research, v. 109, no. E6, doi: 10.1029/2003JE002178.

Wilson, L., and Head, J.W., 1994, Review and analysis of volcanic eruption theory and relationship to observed landforms: Reviews of Geophysics, v. 32, p. 221–264, doi: 10.1029/94RG01113.

Wilson, L., and Mouginis-Mark, P.J., 2003, Phreatomagmatic explosive origin of Hrad Vallis, Mars: Journal of Geophysical Research, v. 108, doi: 10.1029/2002JE001927.

Wood, C.A., and Kienle, J., eds., 1990, Volcanoes of North America, Cambridge, UK, Cambridge Univ. Press, 354 p.

Yang, W., and Ahrens, T.J., 1998, Shock vaporization of anhydrite and global effects of the K/T bolide: Earth and Planetary Science Letters, v. 156, p. 125–140, doi: 10.1016/S0012-821X(98)00006-5.

Zielinski, G.A., Mayewski, P.A., Meeker, L.D., Whitlow, S., Twickler, M.S., Morrison, M.D., Meese, D.A., Gow, A.J., and Alley, R.B., 1994, Record of volcanism since 7000 B.C. from the GISP2 Greenland ice core and implications for the volcano/climate system: Science, v. 264, p. 948–952, doi: 10.1126/science.264.5161.948.

Zielinski, G.A., Mayewski, P.A., Meeker, L.D., Whitlow, S., and Twickler, M.S., 1996a, A 110,000-yr record of explosive volcanism from the GISP2 (Greenland) ice core: Quaternary Research, v. 45, p. 109–118, doi: 10.1006/qres.1996.0013.

Zielinski, G.A., Mayewski, P.A., Meeker, L.D., Whitlow, S., Twickler, M.S., and Taylor, K., 1996b, Potential atmospheric impact of the Toba megaeruption ~71,000 years ago: Geophysical Research Letters, v. 23, p. 837–840, doi: 10.1029/96GL00706.

MANUSCRIPT ACCEPTED BY THE SOCIETY 3 NOVEMBER 2008

Terrestrial subice volcanism: Landform morphology, sequence characteristics, environmental influences, and implications for candidate Mars examples

John L. Smellie*

British Antarctic Survey, High Cross, Madingley Road, Cambridge CB3 0ET, UK

ABSTRACT

The origin and evolution of Mars's inventory of volatile elements is pivotal to a wide range of physical, chemical, geological, and biological issues and concerns. The identification of subglacially erupted volcanoes on Mars suggests that ice sheets existed at high and low latitudes repeatedly over geological time, but the importance of those volcanoes is not just as a simple Boolean climate signal. Like terrestrial subglacially erupted volcanoes, they can potentially yield a more holistic range of paleoenvironmental parameters, including ice thickness, thermal regime, and surface elevation. On Earth, at least nine different types of terrestrial subglacial volcanic successions can be identified using landform characteristics, lithofacies, and sequence architecture. The principal characteristics of each are reviewed in this paper, together with the first empirical comparative analysis of the morphometry of the landforms. All were probably erupted in association with wet-based ice and there are different implications for volcanic landforms erupted under different glacial thermal regimes (polar, subpolar). However, they represent our best sources of information with which to assess Mars analogs, some of which (as on Earth) may have been the source of megascale meltwater outburst floods. Applying the results of this paper to three different morphological types of candidate subglacial volcanoes on Mars indicates that it is difficult to suggest a plausible glaciovolcanic analogy for Mars's *tall cones*; they more closely resemble pyroclastic mounds erupted subaerially or subaqueously, under ice-free conditions. Conversely, Mars's *low-domes* may be very extensive, inflated, subglacial "interface sills" formed under comparatively thick ice of any thermal regime. Finally, the very large, *flat-topped constructs* on Mars resemble mafic tuyas emplaced in thick (up to 2 km) temperate ice. However, because of their very large size compared to terrestrial analogs, the possibility also exists that the latter are polygenetic stratovolcanoes, formed subglacially either within very thick ice, or as multiple superimposed lava-fed deltas emplaced in much thinner ice that repeatedly re-formed on the volcanoes after each eruptive episode. A plausible terrestrial analogy for the latter is the long-lived James Ross Island stratovolcano in Antarctica.

*jlsm@bas.ac.uk

INTRODUCTION

Examples of volcano–ice interactions are common on Earth (e.g., Smellie and Chapman, 2002, and papers therein). Because eruptions have also been observed and extinct edifices are common and amenable to detailed close-up field analysis after ice sheet decay, the internal structure, hydraulic evolution, and eruptive processes are relatively well investigated (e.g., Smellie and Skilling, 1994; Gudmundsson et al., 1997, 2004; Smellie, 2000, 2001, 2006; Tuffen et al., 2002a). Volcanic eruptions through ice sheets also routinely release enormous volumes of meltwater, typically in dramatic and sudden floods known as jökulhlaups. These floods often occur on a terrestrial megascale. For example, even small subglacial eruptions, such as that at Gjálp, Iceland, which erupted just 0.8 km^3 of material (Gudmundsson et al., 2002), resulted in a release of meltwater that was briefly the second largest freshwater discharge of Earth, exceeded only by the Amazon River. Such floods can have a devastating impact on human populations, which may be situated in remote areas and are therefore particularly vulnerable. Glaciovolcanic sequences also represent a valuable, but currently underused, proxy for documenting paleoenvironmental changes over time, and they can be used to obtain critical parameters of past ice sheets (e.g., Wilch and McIntosh, 2002; Mee et al., 2006; Smellie et al., 2006). Finally, the terrestrial examples also provide our best empirical basis for interpretation of remotely observed Martian features.

Reliably identifying and documenting the occurrence of subglacially erupted Mars volcanoes, spatially and over geological time, is central to understanding the inventory and dynamic behavior of past and present water on Mars, and is critical in interpreting correctly the geological imprint of their eruptions. Moreover, it provides a means of investigating the history of past paleoenvironments on a planet other than Earth. The long lifetimes of many Mars volcanoes (tens of millions to hundreds of millions of years: e.g., Neukum et al., 2004) and associated hydrothermal systems may also provide hospitable environments suitable as exobiological refugia. The volcanoes provide a critical link between the deep planet interior, its surface, and biological processes. Therefore understanding the intrinsic and extrinsic environmental controls on subglacial eruptions is an important research frontier, one that is central to our ability to decipher ancient planetary glacial records.

This paper presents the first comprehensive review of terrestrial subglacial edifice and sequence types for both mafic and felsic eruptive systems. It is followed by an empirical assessment of the morphometry of the glaciovolcanic landforms, the most important glacier parameters, and how different glacial thermal regimes may impact on subglacial eruptions. Based on these results, the paper closes with speculations for selected candidate subglacial volcanic edifices on Mars.

MORPHOLOGY, LITHOFACIES, AND SEQUENCE CHARACTERISTICS OF TERRESTRIAL SUBGLACIALLY ERUPTED VOLCANOES

Subglacial eruptions on Earth have formed several different types of volcanic outcrops (e.g., Smellie and Skilling, 1994; Smellie, 2000; Edwards and Russell, 2002; Kelman et al., 2002; Tuffen et al., 2001, 2002a). They show differences in morphology, lithofacies, and sequence architecture. The differences principally reflect variations in magma composition and environmental variables, such as original ice thickness and meltwater hydraulics. A survey of the published literature suggests that the outcrops can be divided into at least nine types, and they are reviewed here for the first time (Table 1; Fig. 1). All but two are volcano edifices; the two exceptions, described in this paper as types of subglacial sheetlike sequences, are not edifices but are volcanic sequences that accumulated away from their edifices; i.e., they are outflow facies. However, they are also a common and distinctive subglacially erupted sequence type, and they are included here for completeness. It is likely that the number of sequence and morphological types will continue to increase as our knowledge of subglacial eruptions improves. Most of the published examples probably formed during relatively short-lived monogenetic eruptions beneath temperate or wet-based ice and the following descriptions and interpretations should be viewed in that context.

In the descriptions that follow, the volcanoes are divided into mafic and felsic types. While the mafic examples are typically basalt or basaltic andesite in composition, and all the more evolved magmas are grouped here as felsic types, it is likely that the distinction in nature is not so clear-cut in respect of the resulting landforms and lithofacies. Thus it is possible that some felsic subglacial volcanoes, particularly of andesitic composition, will be found that show features more commonly found in mafic systems, and vice versa (e.g., Stevenson et al., 2009). For simplicity, this section only considers volcanoes that interacted with unconfined surface ice (i.e., the cryosphere sensu stricto); subsurface interactions with ground ice are ignored, as are snow-contact interactions and effusive products of subaerial eruptions which simply pooled against surface ice ("ice-marginal flows" of Lescinsky and Fink, 2000; see also Kelman et al., 2002; Mee et al., 2006; Stevenson et al., 2006).

Mafic Tuyas

Mafic tuyas were named by Mathews (1947). They are the most distinctive subglacial volcanic edifices and are also known nongenetically as table mountains or mesa landforms, although not all of the latter are tuyas. The edifices are typically small and steep-sided, with flat or gently domed summit regions with or without a crater, and oval to straight-sided plan shapes (van Bemmelen and Rutten, 1955; Jones, 1969; Fig. 2). Although commonly regarded as products of monogenetic eruptions, some described examples are polygenetic, despite their relatively small size, and

TABLE 1. CLASSIFICATION OF TERRESTRIAL SUBGLACIAL LANDFORMS AND SEQUENCES

Landform	Mafic tuya	Felsic tuya (1)	Felsic tuya (2)	Felsic domes and lobes
Subtype		Tephra-dominated	Lava flow dominated	
Composition	Basalt	Andesite, dacite, rhyolite, trachyte, phonolite	Andesite, dacite, rhyolite, trachyte, phonolite	Andesite, dacite, rhyolite, trachyte, phonolite
Morphology	Table mountain, mesa	Table mountain, mesa	Flat-topped column, bladed ridge	Tall steep-sided domes, lobes or irregular
Characteristic lithologies	Subaerial lava (mainly pahoehoe), hyaloclastite, hyalotuff	Sheet lava, ash, pumice, polymict sediments	Sheet lava, ash, pumice, polymict sediments	Lava, glassy breccia
Architecture	Lava-fed delta on pyroclastic cone (tindar)	Lava-capped table mountain; ash core	Lava column (exogenous)	Lava mound (mainly endogenous?)
Deposit type	Vent edifice	Vent edifice	Vent edifice	Vent edifice
Eruptive environment	Ice sheet; glacial lake	Ice sheet	Ice sheet	Alpine glacier, mountain ice cap, ice sheet
Known distribution	Iceland, Canada, Antarctica	Iceland	Canada	Iceland, Canada
Key publications	Mathews (1947); Jones (1969, 1970); Skilling (1994, 2002); Smellie and Hole (1997); Werner and Schmincke (1999); Smellie (2001, 2006)	Tuffen et al. (2002a)	Mathews (1951); Kelman et al. (2002)	Tuffen et al. (2001, 2002b); Edwards and Russell (2002); Kelman et al. (2002)

Landform	Tephra mound/ridge (2 types)	Pillow mound/sheet	Sheetlike sequence (1)	Sheetlike sequence (2)
Subtype	Mafic; felsic		*Mount Pinafore type**	*Dalsheidi type**
Composition	Basalt, andesite, dacite, rhyolite	Basalt	Basalt	Basalt
Morphology	Steep cone or ridge	Low oblate smooth mounds, subdued ridges, extensive sheets	Thin sheets, sinuous ribbons	Thin and thick sheets and sinuous lobes
Characteristic lithologies	Well stratified hyalotuff (mafic); "Massive" hyalotuff, intrusions (felsic)	Pillow and (probably) sheet lava	Hyalotuff sand and gravel, sheet and pillow lava, hyaloclastite, diamict	Intrusive sheet lava, hyaloclastite breccia, diamict
Architecture	Pyroclastic cone	Pillow lava mound/sheet	Esker-like sediments, sheet lava and hyalotuff	Extensive sheet lava ("sill") and cogenetic breccia carapace
Deposit type	Vent edifice	Vent edifice	Outflow facies	Outflow facies
Eruptive environment	Ice sheet; flooded (mafic) or drained (felsic) vault	Ice sheet; not well known	Alpine glacier, mountain ice cap; flowing water	Thick ice sheet
Known distribution	Iceland, Antarctica	Iceland, Canada, Antarctica?	Iceland, Canada, Antarctica	Iceland, Canada, Antarctica
Key publications	Skilling (1994); Smellie and Hole (1997); Werner and Schmincke (1999); Smellie (2001)	Smellie and Hole (1997); Dixon et al. (2002)	Smellie et al. (1993); Loughlin (2002)	Walker and Blake (1966); Bergh and Sigvaldason (1998); Loughlin (2002)

*Provisional names.

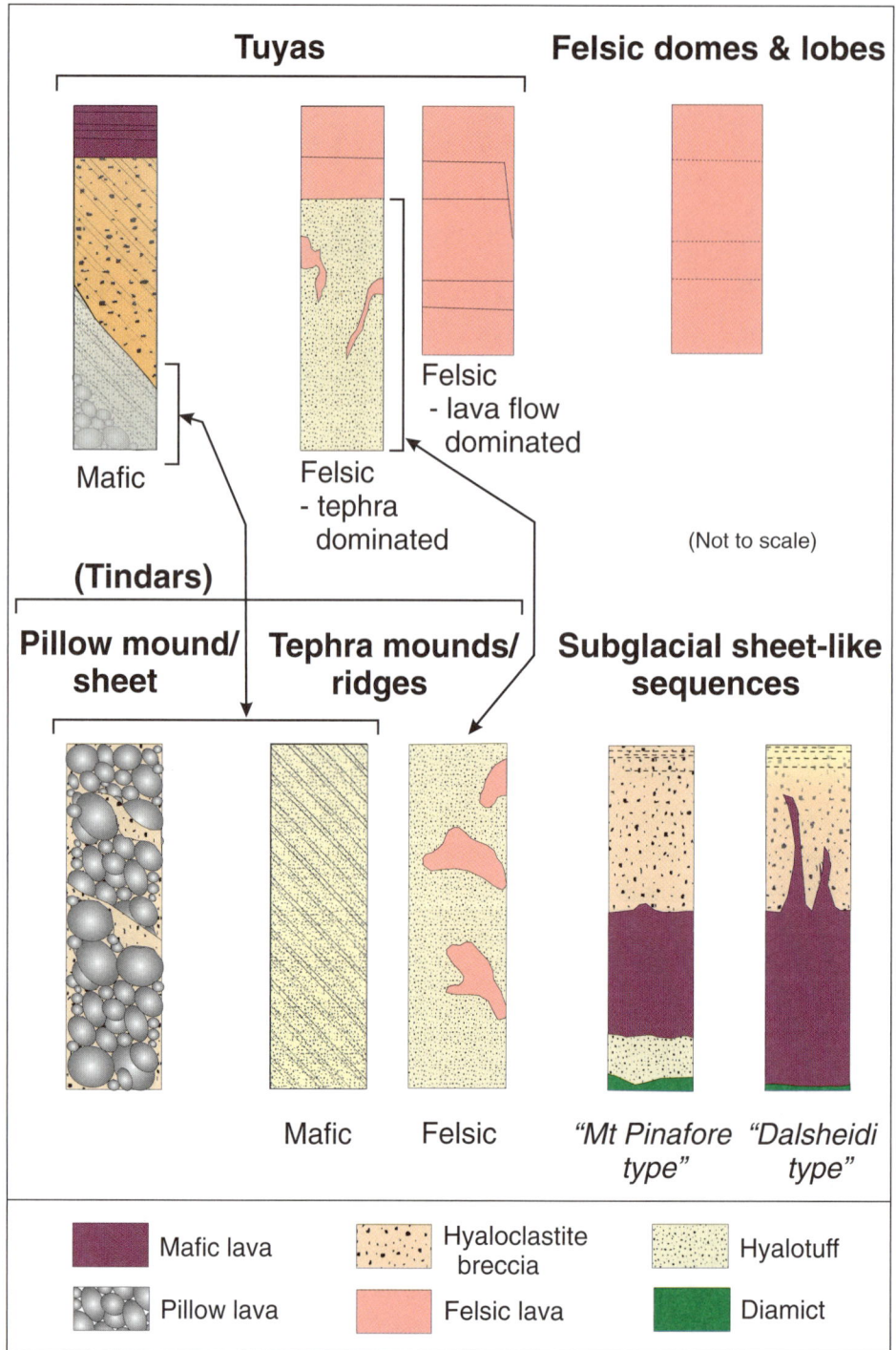

Figure 1. Schematic vertical profile sections through the nine different subglacial landform types identified and described in this paper, illustrating the principal lithofacies and sequence architecture.

Figure 2. View of a mafic tuya, showing characteristic mesalike summit and very steep flanks. The tuya is ~600 m high and 4 km in length. Although showing features of a classic monogenetic edifice, this landform is polygenetic and underwent a late resumption of activity, creating a thin lava-fed delta on the larger, older delta surface. Hlodufell, Iceland. Photograph by the author.

a monogenetic history should not always be assumed (Jones, 1969, 1970; Werner and Schmincke, 1999; Smellie, 2001). Published examples of much larger polygenetic tuya edifices are scarce. They have low surface profiles more akin to shield volcanoes, but retain the steep margins characteristic of confinement by ice during eruption(s) (Jones, 1970). The largest polygenetic terrestrial example known to the author is the Mount Haddington volcano, on James Ross Island, Antarctica. It rises 1.5 km, has a basal diameter of 60 km, and encircling cliffs 600 m high. It was constructed in multiple extrusive phases over a period of at least 6.5 Ma (probably 10 Ma) since the latest Miocene (Nelson, 1975; Smellie, 1999; Smellie et al., 2008).

Mafic tuyas are *defined* by the presence of lava-fed deltas (sensu Skilling, 2002; Figs. 1, 3). The deltas represent a late subaerial progradational phase of edifice-building (Smellie, 2000) and they have distinctive sequence architecture (Fig. 3). Although the deltas are constructed on top of a subaqueous tuff cone and/or pillow mound (see later), the earlier-formed constructs do not significantly affect the final shape of the edifice. The flat or gently convex top is a subaerial shield volcano formed of lavas with a summit crater or small pyroclastic cone of scoria or agglutinate. In a lava-fed delta, the lavas are the volcanic equivalent of topset beds in a sedimentary delta. They overlie cogenetic coarse hyaloclastite breccias (equivalent to sedimentary delta foreset beds), which dip radially outward at steep repose angles (~25°–40°). Although, by back-projecting enough dip directions, the approximate location of the source vent may be identified, in practice volcanic deltas advance as lobes and the bed orientations are highly variable (cf. fig. 3 in Skilling, 2002). Thus locating the likely source location is often not a trivial exercise and may be impossible. The breccias are formed by a combination of quench-induced shattering and avalanching at the delta brink point (Skilling, 2002). The junction between the lavas and dipping breccias is a planar surface known as a passage zone, which records the water level coeval with delta formation (Jones, 1969, 1970; Smellie, 2006). The passage zones usually appear to be horizontal surfaces but careful mapping has demonstrated that they can maintain a significant dip (a few degrees) over distances of several km. New work has shown that passage zone elevations can also vary through several tens of meters, up and down, over quite short horizontal distances of a few hundred meters (Smellie, 2006). Original gentle dips and rapid elevation variations in passage zones are unlikely to form in marine or nonglacial lacustrine environments and such attributes are diagnostic of a glacial environment, in which the contemporaneous ice-impounded water level can vary significantly. Such observations are highly important in identifying an eruptive paleoenvironment.

The principal limitation on lateral extent of a mafic tuya is the volume of magma discharged over time, since deltas will advance as long as space is available and magma is supplied. By contrast, tuyas formed by eruptions within wet-based temperate glaciers could be height-limited. This suggestion is speculative and follows from the modeling by Smellie (2000, 2001). It would occur if the water level in lakes associated with mafic tuyas was pinned by the elevation of the meltwater overflow point (spillway), an assumption of that modeling. To prevent glacier flotation and associated sudden catastrophic basal meltwater drainage, the glacier must contain a relatively thick capping layer of permeable firn and/or fractured (crevassed) ice, through which the meltwater could overflow. The maximum thickness of such a permeable layer is limited to ~150 m on Earth, and is typically only a few tens of meters for temperate systems. Thus, beyond a critical thickness for the glacier body, the overlying permeable layer is too thin and the lake surface will not reach it before the glacier

Figure 3. Lava-fed delta ~200 m thick, showing distinctive sequence architecture, comprising flat-lying subaerial pahoehoe caprock overlying homoclinal steep-dipping hyaloclastite breccias, representing top-set and fore-set beds, respectively. The prominent planar surface separating lavas from breccias is a passage zone (a "fossil" water level). James Ross Island, Antarctica. Photograph by the author.

is floated. The limiting thickness for glaciers is ~750–1000 m, above which the pressure achieved at the base of a glacial lake will always be enough to float the glacier. However, this conclusion is based on an assumption that meltwater escape is dominantly by overflowing. The dominant mode of discharge during tuya eruptions, whether basal or supraglacial, is still unclear; both may occur (cf. Gudmundsson et al., 1997; Smellie, 2006). Thus in a tuya dominated by high basal meltwater discharge, the water in the associated lake may never reach a depth sufficient to float the glacier (i.e., ~90% of the ice sheet thickness). Thus water depths exceeding 750–1000 m may be achieved, leading to correspondingly taller edifices.

Felsic Tuyas

It appears that felsic magmas do not form lava-fed deltas. However, some subglacial flat-topped, steep-sided landforms with felsic compositions have been described as tuyas (Kelman et al., 2002; Tuffen et al., 2002a; cf. Mathews, 1951). Compared to mafic tuyas, few felsic examples have been described. Their compositions include andesite, dacite, and rhyolite. The internal structure of felsic tuyas is very different from mafic tuyas, and at least two types are recognized and identified here as (1) tephra-dominated and (2) lava flow dominated (Kelman et al., 2002, fig. 4; Tuffen et al., 2002a, fig. 3).

In *tephra-dominated felsic tuyas*, the flat cap is formed of horizontal or gently dipping, relatively thin (typically a few tens of meters) rhyolite lavas showing prominent flow foliation, fine marginal columnar joints, and subaerial flow top breccias. The lavas overlie a much thicker pile (a few hundred meters) of fine phreatomagmatic tephra, which is typically obscured by lava scree and therefore poorly exposed (Tuffen et al., 2002a). Although typically massive in appearance, the tephra is often faintly stratified. The deposits also locally preserve coarsely stratified polymict fragmental rocks (volcaniclastic) deposited from flowing water. The latter indicate meltwater drainage during eruption, broadly similar to the mode of occurrence of basal sediments in mafic tuyas described by Smellie (2001) but in at least one case in a higher stratigraphic position. The relationships were described as consistent with rhyolite magma emplaced in an englacial vault that was drained throughout eruption (Tuffen et al., 2002a). With the relatively thin ice thicknesses envisaged for the few edifices described so far (> 350 m), the naturally higher volatile contents expected in these felsic magmas may simply result in violently explosive subaerial phreatomagmatic eruptions in a water-rich vault from the earliest stages (cf. Tuffen et al., 2007).

By contrast, *lava flow-dominated felsic tuyas*, as their name suggests, are distinctive flat-topped columns or short ridges formed of flat-lying felsic lavas (Fig. 4). The lavas show marginal water-cooling and subaerial flow tops similar to the capping lavas in tephra-dominated felsic tuyas. Only andesite–dacite systems have been described so far. In some, lavas are also plastered onto the subvertical tuya flanks and are believed to have been emplaced by extruding down coeval "bergschrunds" (Mathews, 1951). Fragmental deposits are practically absent, other than rare basal nonvolcanic diamict (Mathews, 1951; Kelman et al., 2002). The absence of volcaniclastic rocks and, by inference, explosive water–magma interactions, was explained as a consequence of lower total heat available and less efficient heat exchange, resulting in positive cavity pressures during eruption, and essentially dry vents perched at high elevations and surrounded by steep slopes (Kelman et al., 2002; see also Tuffen et al., 2001, 2007). These would cause meltwater to drain efficiently from the cavity. However, effusion of gas-poor magma is probably important for constructing these types of tuya, otherwise tephra-dominated felsic tuyas might be expected. Although the presence of a substantial thickness of ice overburden may also suppress explosive magmatic vesiculation, it would not promote the development of subaerial lava flow tops.

Pillow Mounds, Ridges, and Sheets

Pillow mounds and sheets are formed of pillow lava and hyaloclastite breccia. The breccia is thought to have formed mainly as mass flows derived by gravitational collapse of the associated pillow lava pile (cf. Furnes and Fridleifsson, 1979). All described examples are basaltic in composition. Many occur in the cores of mafic tuyas, where pillow lava effusion is usually regarded as a ubiquitous early stage of tuya construction (e.g., Jones, 1969, 1970; Skilling, 1994; Smellie and Hole, 1997). This assumption has been questioned recently for outcrops where pillow lava may be absent (Gudmundsson et al., 2004; LeMasurier, 2002), but examples are rare (e.g., Werner and Schmincke, 1999, Fig. 3). It is an important issue, however, since it influences our understanding of how heat is transferred from magma to melt ice. Edifices with central vents and formed solely of pillows and breccia form relatively low oblate mounds with moderately dipping sides and smooth profiles (Fig. 5), while those erupted from fissures form subdued ridges with gentle slopes. Examples may include the subglacial volcanoes in British Columbia examined by Dixon et al. (2002), which are formed solely of pillow lava extruded under ice sheets that may have been > 2 km thick. By contrast, pillow sheets are only a few tens of meters thick yet may extend laterally 10 km or more (Snorrason and Vilmundardottir, 2000). Sheets are currently only known from Iceland. They are poorly described and understood so far. Evidence for syneruptive confinement of a pillow mound includes plastic molding of pillows against a former ice wall in an Iceland tuya (I. P. Skilling, 2002, personal commun.). Such banking may lead to over-steepening of a pillow mound compared with submarine examples, with which the subglacial mounds are otherwise indistinguishable (cf. Staudigel and Schmincke, 1984).

Smellie (2000) postulated, on theoretical grounds, that some monogenetic pillow mounds may be draped by lava and/or pyroclastic deposits formed subaerially at a late eruptive stage. Under relatively thick glaciers (> several hundred meters) formed entirely of ice or with a very thin permeable upper layer, meltwater that accumulates in a vault to > 90% of the thickness of the ice sheet will always float the enclosing ice and release a sudden flood (jökulhlaup). Since a stable lake is necessary for lava-fed deltas to form, construction of a tuya is unlikely (Smellie, 2000, 2001). Following vault drainage, subaerial volcanic lithofacies will be erupted if the eruption continues. Thus the juxtaposition of *cogenetic* subaqueous and subaerial lithofacies in this asso-

Figure 4. View of the Table, British Columbia, Canada. Described by Mathews (1951), this is the best-known example of a lava flow-dominated felsic tuya. Photograph by Cathie Hickson.

ciation is probably diagnostic of extrusion under wet-based ice sheets several hundred meters thick and formed almost entirely of ice (s.s.) (Smellie, 2000). The only published examples known to the author are pillow lava centers draped unconformably by agglutinate at Seal Nunataks, Antarctica, which formed within an ice sheet > 600 m thick (Smellie and Hole, 1997). Note that pillow mounds lacking the unconformable subaerial draping lithofacies may be impossible to distinguish from pillow mounds in volcanoes erupted under much thinner glaciers, and which simply ceased activity prematurely.

Figure 5. Pillow mound south of Reykjavik, Iceland. The mound was erupted from a central source and has a low oblate shape and smooth profile (delineated by the white line) broken only by a summit tephra mound (paler gray shaded region) formed at a late stage in the eruption. Photograph by the author.

Tephra Mounds and Ridges

Tephra mounds (the volcanic mounds of Hickson [2000]) and ridges are known for subglacial volcanoes of mafic and felsic compositions. Mafic examples are by far the most common. They are comparatively steep-sided and, when erupted from fissures, form prominent ridges with multiple summit peaks. They were called *tindars* by Jones (1969), a useful but etymologically incorrect term that encompasses not only tephra ridges or mounds, but also their pillow lava cores (Smellie, 2000). Two types can be distinguished based on sequence details; there are also morphometric differences (see later).

Mafic tephra mounds are always prominently stratified, comprising relatively thin beds of hyalotuff (sensu Honnorez and Kirst, 1975) showing a variety of planar bedforms and sedimentary structures characteristic of unconfined sediment gravity flows (Skilling, 1994; Smellie and Hole, 1997; Werner and Schmincke, 1999; Smellie, 2000, 2001). They are formed exclusively of sand- and gravel-sized, variably vesiculated volcanic glass, and represent redeposited syneruptive tephra deposits of a subaqueous tuff cone. The dominance of sediment gravity flow deposits over fall deposits in these tuff cones is a characteristic feature of shallow water explosive eruptions and is a principal consequence of an eruption plume interacting with a water column (e.g., White et al., 2003). Stratification is mainly planar and continuous, and dips rapidly flatten out at higher positions in the tephra pile. Uniquely also, higher beds may dip back to source (Smellie, 2000, fig. 5).

Felsic tephra mounds and ridges are largely undescribed so far, except where they occur below a subaerial lava cap in tephra-dominated felsic (rhyolite) tuyas (Tuffen et al., 2002a). They also occur without a lava caprock (e.g., in the Kerlingarfjoll rhyolite complex, central Iceland; Smellie, J. Stevenson, and D. McGarvie, 2002, personal observ.). They form relatively steep-sided pale-colored successions, commonly with numerous, sometimes voluminous syneruptive rhyolite hypabyssal intrusions. Stratification is uncommon and typically hard to discern. The tephra vary from mainly fine-grained felsic hyalotuff characteristic of phreatomagmatic eruptions to less common coarser pumice-rich deposits similar to Plinian successions. Dense obsidian bombs or blocks are common.

All the published evidence for mafic tephra mounds and ridges indicates rapid upward edifice aggradation within ponded water, which is a glacial lake in this context. Uniquely so far, one mafic outcrop shows a thin sequence of basal sediments with unidirectional traction current structures, believed to have been deposited by water currents generated at the base of an ice sheet by water flushed from an englacial vault (Smellie, 2001). This is the only published lithofacies evidence for a leaky basaltic englacial vault (Smellie, 2000) although similar evidence has been described to the author for the basaltic tuya at Hlodufell, Iceland (I. P. Skilling, 2002, personal commun.), and basal meltwater escape was spectacularly demonstrated throughout the 1996 eruption at Gjálp, Iceland (Gudmundsson et al., 1997). The hyalotuff stratification rapidly flattens out radially away from the vent. In some cases, beds are even back-tilted toward the source, signifying that they were deposited in a narrow basin confined by ice walls now melted away, and which was rapidly filled by tephra (Smellie, 2000). Tephra mounds and ridges may have flanks with significantly steeper-than-repose slopes than would be achieved in subaqueous tuff cones constructed in laterally unconfined nonglacial environments (lacustrine, marine). Intuitively this may be a consequence of their ice confinement (Walker and Blake, 1966; Werner and Schmincke, 1999), rapid diagenesis and lithification (pore water temperatures may be as high as 150–300 °C in the glassy pile; Gudmundsson, 2003), and early oversteepening of the lower flanks of the indurated edifice by glaciers after eruptions have ceased. In a pristine state, the steep flank faces should reveal only gently dipping bedding, both slope-parallel and locally flat to back-tilted where deposits were ponded against the former ice walls. However, tuff cone flanks will commonly be regraded by gravitational collapse(s) as the ice walls melt back during eruption and remove their support. Thus syneruption collapse scars are a conspicuous and characteristic feature, seen in dip and strike faces as large-scale concave-up surfaces draped by only slightly younger beds (sketched in Skilling, 1994; Smellie and Hole, 1997; see also Smellie [2001] for other collapse-related features).

By analogy with tephra cores in felsic tuyas, the felsic mounds probably also formed by explosive phreatomagmatic eruptions in wet vaults (see above). Eruption-related extreme turbulence and associated percussive effects may be of higher intensity in the more gas-rich viscous felsic systems compared with mafic magmas. Such effects may cause repeated large-scale dewatering, collapse, and homogenization of large parts of the tephra pile, thus destroying much of the evidence for stratification.

Felsic Lobes and Domes

Several steep-sided subglacially erupted andesite, dacite, rhyolite, and trachyte–phonolite flows with fine-scale columnar jointing have been described from British Columbia and Iceland (Tuffen et al., 2001, 2002b; Edwards and Russell, 2002; Kelman et al., 2002). Original conical, lobe, domelike, or irregular outcrop shapes are evident from the radiating patterns of the columnar cooling joints and presence of prominent thick glassy chilled surfaces. The joint patterns and chilled surfaces are caused by ice contact and record the three-dimensional orientation of the local cooling front. Associated fragmental lithofacies are usually uncommon, but subglacial domes in Canada are encased in patchy monomict glassy breccia, which may be "agglutinated" (sintered) where overridden. Associated lithofacies in an Icelandic example are more variable and include a variety of polymict meltout tills and monomict sands and gravels washed into subglacial cavities by traction currents (cf. mafic sheetlike sequences; below), together with variably perlitized ashy-matrixed breccias formed by quenching and in situ granulation of rhyolite flow lobes. The Icelandic outcrop also shows flow lobes and associated breccias remobilized in hot and cold mass flows.

Formation of these structures is probably a result of relatively slow effusion of degassed felsic magma into a drained and thus essentially dry vault (cf. rationale used by Tuffen et al., 2002a, after Hoskuldsson and Sparks, 1997). They might be regarded as equivalent to the earliest-formed extrusive layers within the much taller lava flow-dominated felsic tuyas. The difference in edifice morphology between felsic domes/lobes and flow-dominated tuyas may be due to endogenic growth dominating in the former, yielding domes or lobes rather than lava piles. The shape of a subglacial dome probably reflects that of an enclosing ice vault, which remained roofed-over (cf. Tuffen et al., 2001, 2002b).

Subglacial Sheetlike Sequences

Subglacial sheetlike sequences are relatively thin deposits composed of different combinations of four main lithofacies (below). Also called subglacial sheet flow sequences (Smellie, 2001, p. 10), they were renamed as subglacial sheetlike sequences using a nongenetic terminology (Smellie, 2008). Practically all published examples are basaltic, but Edwards and Russell (2002) have described a few trachyte and phonolite sequences in British Columbia. Two different sequence types can be distinguished, informally named the Mount Pinafore and Dalsheidi types (Table 1; Fig. 1). Distinction is based on rather subtle but genetically important differences in lithofacies and sequence architecture, which have significant implications for interpreting original ice sheet thicknesses. Both correspond to sequences of outflow facies. They are thus not volcanic edifices, but form a distinctive subglacial volcanic sequence type nevertheless. Formal characterization and interpretation of the origin of these sequences are still at an early stage (Smellie, 2008).

Deposits of the first sequence type are referred to collectively as sheetlike sequences of *Mount Pinafore type* (Smellie, 2008). They are usually only a few tens of meters thick. There is considerable variability in the relationships between the various lithofacies and a wide range of examples has been described in Iceland, Antarctica, and British Columbia (Carsewell, 1983; Smellie et al., 1993; Kelman et al., 2002; Loughlin, 2002). Those in Antarctica, at Mount Pinafore, were defined as sequence holotypes for this kind of eruption (Smellie and Skilling, 1994). The sequences are commonly repetitive and they are usually topographically confined between sharply defined glacially modified (polished, striated, molded) surfaces (e.g., Smellie et al., 1993; Kelman et al., 2002; Loughlin, 2002). The individual eruption-generated units have been described as eskerlike (Mathews, 1958; Smellie and Skilling, 1994). They commonly form narrow, sinuous, anastomosing low outcrops but lobe or sheet outcrops also occur. They sometimes have steep flanks thought to be due to banking against former ice walls (Loughlin, 2002). However, the profile, overall shape, and dimensions of an outcrop will depend on the topography of the underlying bedrock (e.g., incised valley), erupted volume of magma, and discharge rate.

Outcrops typically commence in glacial diamict (representing melted-out glacial bedload) and pass up abruptly into volcaniclastic sediments (monomict vitroclastic gravels and sands representing redeposited hyaloclastite and hyalotuff). The volcaniclastic deposits show stratification and other features generated by traction currents, and they are a consequence of transport within and deposition from flowing water, often in channels. Some beds are likely hyperconcentrated-flow deposits (Smellie et al., 1993; Loughlin, 2002). Sheet lava, locally pillowed and with conspicuous entablature or kubbaberg cooling joints caused by water chilling, overlies the hyalotuffs (kubbaberg is an Icelandic termed meaning cube-jointed [water-chilled] basalt; Bergh and Sigvaldason, 1991). Rare examples contain glassy chilled and radial-jointed basal cavities up to several meters across probably caused by melting out of blocks of ice engulfed by the lavas during emplacement (Walker and Blake, 1966; Loughlin, 2002). The lava is encased in and overlain by massive hyaloclastite breccia. The latter has two principal origins: by brecciation and mechanical spallation essentially in situ; and as laterally extensive mass flow deposits which may transform upward into crudely planar stratified hyperconcentrated-flow deposits that may contain admixed hyalotuff debris. The outcrops rarely pass up into subaerial, essentially dry eruptive products (lava and/or oxidized scoria, locally welded), as has been observed in Iceland (Loughlin, 2002) and Antarctica (Smellie, 1990, personal observ.). The presence of basal explosively generated hyalotuffs, abundant evidence for transport by flowing water, water-chilled lava, and occurrence in thin sequences, were interpreted by Smellie et al. (1993) as indicating eruptions beneath thin glaciers (< ~ 150 m) mainly formed of permeable firn, snow, and fractured ice. Deposition may have been in subglacial tunnels. Alternatively, rapid disintegration of a thin glacier overlying tunnels may also occur, with deposition in ice-flanked subaerial channels. Volcanic edifices responsible for the Mount Pinafore-type sequences are very poorly known. Those described are simple pyroclastic cones formed during subaerial magmatic and phreatomagmatic eruptions (Smellie et al., 1993; Smellie, 2002). They are unremarkable and undiagnostic, in themselves, of a glacial environment.

Deposits of the second type are much less well known. They are called subglacial sheetlike sequences of *Dalsheidi type* in deference to the earliest substantive description by Walker and Blake (1966; cf. Smellie, 2008). Dalsheidi-type sequences can extend to much greater thicknesses than Mount Pinafore types. For example, the Dalsheidi outcrop itself, in southeast Iceland, is > 20 km long and comprises a basal lava up to 100 m thick in a breccia-dominated deposit that may have an aggregate thickness of 300 m. Another outcrop with comparable lithofacies and dimensions, but up to 5 km wide, is also present near Katla (K. Sæmundsson, 2003, personal commun.). Larsen (2002) described a similar but smaller outcrop nearby ("Kriki hyaloclastite flow"), which emerges from beneath the Kötlujökull glacier; it possibly has a very young age (tenth century? Larsen, 2002). Loughlin (2002) also reported numerous lobate outcrops at Eyjafjallajökull, which are individually several km wide and long and vary in thickness from 5 to 150 m thick. The sequences described by these authors consist of three major volcanic subunits or lithofacies, comprising: (1) basal sheet lava tens to hundreds of meters in width, locally pillowed, and typically showing colonnade and cube-jointed entablature (kubbaberg). The lava has a sharp base but its upper surface displays multiple irregular flamelike apophyses, lobes, and dykelike masses that intrude, bud, and spall into (2) overlying massive hyaloclastite breccia and (3) an upper unit of horizontally bedded gravel-grade hyaloclastite breccia. Highly vesicular hyaloclasts may be mixed with the uppermost gravel breccia deposits. The thinned lateral margins of the lava sheet commonly curve up into the hyaloclastite breccias and gravels. Both Larsen (2002) and Loughlin (2002) noted similarities in lithofacies to extensive deposits in Iceland described by Bergh and Sigvaldason (1998), to which Bergh and Sigvaldason ascribed a marine origin that relied on assuming an uncommonly high relative sea level and interpretation of the associated sedimentary rocks. Those similarities were emphasized by Smellie (2008) and a subglacial environment postulated.

The lithofacies and sequence architecture of the Dalsheidi-type deposits are similar to the thinner Mount Pinafore-type deposits (Fig. 1), but the outcrop dimensions are typically (but not always) greater. More importantly, there are also significant differences in lithofacies and sequence architecture and different environmental conditions are implied: (i) there is neither a basal unit of hyalotuffs nor evidence for abundant flowing water in the earliest stages of these eruptions; (ii) the basal sheet lava shows extensive evidence for having intruded its breccia carapace; (iii) the breccia carapace was rapidly emplaced in a laterally widespread hyperconcentrated flood event, with some reworking in the late-stage waning stages (Loughlin, 2002); and (iv) none of the Dalsheidi-type sequences passes up into subaerial lavas. The absence of early formed hyalotuffs suggests that initial eruptions were not explosive and thus were probably damped by a greater ice overburden (> 300 m? cf. Wilson and Head, 2002) than characterized eruptions of Mount Pinafore type. Conversely, the presence of minor highly vesiculated sideromelane in the latest-formed deposits suggests that the eruption became explosive in the latest stages, possibly signifying reduced ice overburden pressures and subaerial eruption in an ice sheet melted through completely. The basal lava clearly intruded its breccia carapace, to which it also contributed by pillow budding and spalling. The breccia is hyaloclastite with an abundance of poorly to nonvesicular sideromelane, and was not formed explosively.

These observations were reconciled within an empirical model of subglacial eruptions beneath a comparatively thick glacier (> 300 m, probably > 1000 m), one largely formed of impermeable ice (Smellie, 2008). Modeling by Wilson and Head (2002) suggested that basalt magma may be injected rapidly along a bedrock–ice interface as thin sheets (sills), which Smellie (2008) equated with the basal lava unit in Dalsheidi-type sequences. Because the rate of diffusive heat loss from sills is much less than their emplacement time, the lateral extent reached by the sill is determined mainly by discharge volume and rate. Emplacement is followed by sill inflation and creation of a meltwater cupola along the sill surface, particularly in the area of highest heat flow, above the vent, where an expanded, water-filled vault will develop. In a prolonged eruption, a pillow pile may construct an edifice over the vent. If the hydraulic pressure in the vault floats the glacier, meltwater will escape as a voluminous and sudden flood (jökulhlaup). The floodwater is capable of eroding and entraining large quantities of pillows and unconsolidated hyaloclastite breccia from the edifice. This is envisaged as the likely source of much of the massive to bedded pillow-fragment hyaloclastite breccia above the earlier-emplaced basal lava (sill). If the sill retains a molten interior, some lava may escape to back-inject the overlying breccia. The pressure drop over the vent during the jökulhlaup may also cause a switch to explosive eruption and generation of hyalotuff, which could then be reworked through the subglacial system in the upper beds of mixed hyaloclastite and hyalotuff sideromelane.

MORPHOMETRY OF TERRESTRIAL SUBGLACIAL VOLCANIC EDIFICES

Having identified the major landform types constructed during subglacial eruptions, it is now possible to examine their morphometry. In this paper, area, volume, and aspect ratio are used to compare the morphometric characteristics. Aspect ratio is defined as the ratio of maximum height (above volcano base) to edifice width. For sheetlike sequences, deposit *thickness* is used rather than height since edifice dimensions are unknown or undiagnostic. Although aspect ratio is a crude parameter and it may differ in detail from true volcano profiles, it is still a useful empirical and informative measure. For example, aspect ratio ignores the presence of summit craters. However, craters, particularly in tephra cones, are often amongst the first casualties of posteruptive erosion, and they are absent in many pre-Quaternary examples.

Area is plotted against volume in Figure 6. The data for most of the landforms fall within a box bounded by 10 km^2 and 2.0 km^3. However, most of the mafic tuyas fall outside of that box, and they generally have the highest volumes of any subglacial volcanic landform. Tuyas, both felsic and mafic, have the highest volume-to-area ratios and define a well-defined linear regression trend. Much lower ratios and a very different regression line characterize tephra mounds/ridges. The data distribution for the latter is very similar to those for subaqueous tuff cones erupted in unconfined, nonglacial settings (see comments later). Data for pillow mounds/sheets and sheetlike sequences are very scattered. They generally have the lowest ratios in the data set corresponding to the highest areas, a reflection of their sheet or ribbonlike morphologies, but very few data are currently available; there are even fewer data for felsic domes.

Figure 7A shows aspect ratios using, as horizontal dimension, the *maximum* basal length of edifices. Maximum horizontal extent is strongly affected by mode of eruption, whether from fissures or point sources. Fissure length is a tectonically controlled variable and not an intrinsic volcano-determined dimension. The different types of volcano outcrops described in this paper generally define discrete fields in the diagram. Overlap is greatest for data from felsic centers (felsic tuyas and domes). Although the data set is still small, the felsic flow-dominated tuyas have the highest aspect ratios measured so far. By contrast, the mafic pillow mounds and sheets, and subglacial sheetlike sequences, have very low aspect ratios unmatched by any other type. In the center of the diagram, data for the tephra mounds/ridges appear to plot on different trajectories to most mafic tuyas. Two tephra mound subfields are also defined compositionally, corresponding to mafic and felsic tephra mounds, although there are too few data to be definitive. Tuya landforms in subfield "a" reflect point

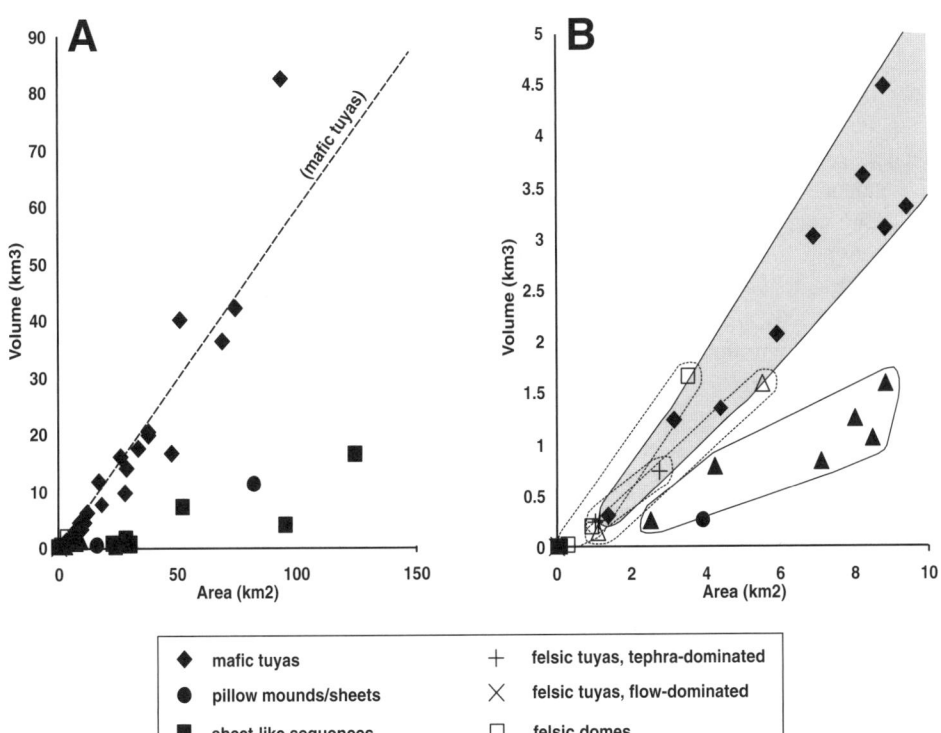

Figure 6. Diagram showing volume (km^3) plotted against area (km^2) for the terrestrial subglacial volcanic sequences described in this paper. A—Full data set. B—Expanded view of data for the smaller edifices, which are the most common types; the field for mafic tuyas is shaded. The areas and volumes of most of the felsic examples are too small to show up clearly in either diagram. See text for discussion.

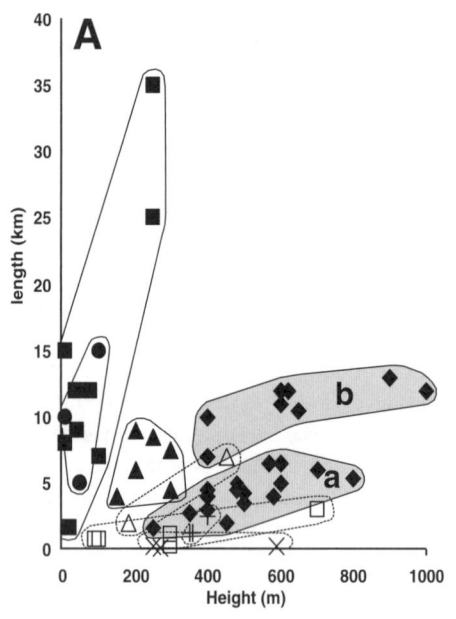
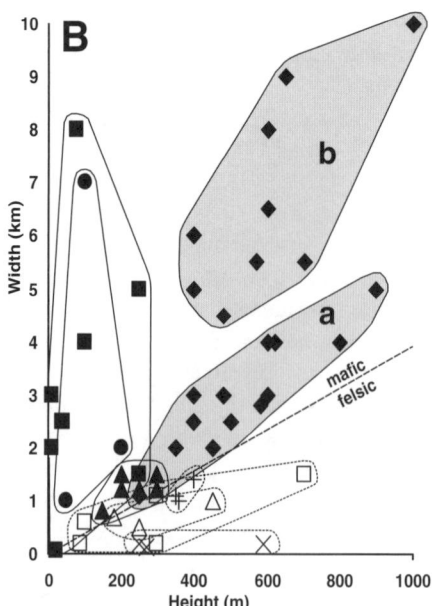

Figure 7. Diagrams showing aspect ratios (height vs. width) for terrestrial subglacial volcanic sequences. In A, height is plotted against maximum basal length, which in many instances (particularly tephra mounds/ridges) is related to constructs formed during fissure eruptions and is strongly controlled by tectonics. In B, height is plotted against minimum basal dimension (width), a parameter that more closely reflects volcanic processes. Fields for mafic tuyas are shaded. See text for discussion.

source eruptions, while those in subfield "b" are elongate fissure-erupted tuyas. That none of the latter examples also falls within the analogous subfield ("b") in Figure 7B is proof that they are not simply products of laterally extensive eruptions from central vents. Comparative use of both of these diagrams discriminates fissure-erupted from large-volume centrally erupted mafic tuyas.

Figure 7B is similar to Figure 7A except that *minimum* horizontal edifice dimensions (widths) are now used as the ordinate. Because width is perpendicular to fissure length, it is likely to reflect volcano construction processes rather than tectonics. The diagram illustrates a clear distinction into mafic and felsic centers, with the highest ratios shown by the latter. Like Figure 7A, several of the different subglacial volcanic sequence types form discrete fields, and there is a better separation into mafic and felsic types across an aspect ratio of ~ 0.25. Although data are still very few, this observation apparently includes tephra mounds and ridges, with the few felsic examples showing significantly higher ratios than their mafic counterparts. The reasons for two fields for mafic and felsic tephra mounds/ridges are uncertain. However, the presence of abundant coeval hypabyssal intrusions in felsic tephra mounds/ridges (Fig. 5) may help to consolidate the erupted piles and stabilize slopes more rapidly, allowing them to retain very high gradients after flank collapses. Interestingly, aspect ratios for mafic tuff cones from nonglacial subaqueous settings are largely indistinguishable from the data for subglacial centers, while extending to much lower ratios (down to 0.051). Thus the prediction that glacially confined tephra mounds should generally have steeper flanks and higher aspect ratios is only weakly suggested by the data available so far, and tuff cones formed in either setting may be indistinguishable. The highest aspect ratios are shown by felsic flow-dominated tuyas, which have a mean value of ~ 3. These landforms are flat-topped columns and short, flat-topped "bladed" ridges. The lowest ratios by far are shown by mafic pillow mounds/sheets and sheetlike sequences. The very wide scatter of aspect ratios for sheetlike sequences (0.003–0.4) may simply be a consequence of some of those sequences being topographically confined. Thus the width of the individual sequences is controlled by nonvolcanic factors.

Most of the tuya data are dispersed around a fairly constant aspect ratio of 0.18. This is comparable to the ratio for most tuyas using the edifice lengths (Fig. 7A, subfield "a"), implying that growth patterns of most mafic tuyas are related to eruptions from central sources rather than fissures, at least in the present data set. The simplest explanation for the subordinate subfield of much wider mafic tuyas ("b" in Fig. 7B) is that those edifices were constructed during unusually voluminous eruptions, when the associated lava-fed deltas simply prograded laterally to a greater extent. Thus the presence of a detached subfield is probably an artifact of a small data set, and the two fields will probably merge as more data become available. The largest mafic tuya measured so far is ~1000 m in height, recording eruption through a contemporaneous ice sheet of comparable thickness. Interestingly, that height is within the range of maximum heights (750–1000 m) predicted for tuyas erupted in temperate ice and dominated by overflowing meltwater hydraulics, discussed above.

DISCUSSION

The Influence of Ice Sheet Properties on Terrestrial Subglacial Eruptions

An important effect of an enclosing ice sheet on an erupting volcano is to confine the resulting products, whether meltwater or magmatic materials. Otherwise, eruptions will be indistinguishable in most other respects from their submarine or lacustrine equivalents (Walker and Blake, 1966). In addition to the simple presence and thickness of a glacial cover, three properties of glaciers are paramount in their influence on subglacial eruptions.

Thermal Regime

The temperature distribution, or thermal regime, of a glacier is fundamentally important for the way a glacier deforms (i.e., rheology), and with regard to the role of meltwater (i.e., hydraulics). Glaciers may be classified empirically by their thermal regime. There are three main types, although transitions may occur in the same glacier as in the present Antarctic Ice Sheet (Wilch and Hughes, 2000). Temperate glaciers are relatively warm, i.e., at the melting point throughout except for a thin surface layer. A thin layer of meltwater intervenes between the glacier sole and underlying bedrock and the glacier moves by sliding. Polar glaciers are much colder and are well below freezing throughout the year. They are frozen to their bed and movement is usually assumed to be accommodated by shear within the glacier itself. Finally, polythermal or subpolar glaciers are also cold glaciers with temperatures well below freezing, but their bases are warmed to the melting point from beneath by geothermal heat and they are therefore locally wet-based. Alternatively, glaciers can be viewed as two main types: wet-based (i.e., temperate and subpolar glaciers) and dry-based (i.e., frozen to their bed; polar glaciers). A major consequence of the thermal regime is that colder glaciers require significantly more energy to melt compared with warmer glaciers, since energy has to be used to warm the cold ice to its melting point. For example, every kg of ice at –50 °C at atmospheric pressure requires 209 kJ of energy simply to raise its temperature to the freezing point. The presence of impurities, such as air, salts, and CO_2, lowers the melting point relative to pure water ice. Soluble and insoluble impurities can greatly increase the effective specific heat capacity of ice.

Rheology

Ice below its pressure melting point is harder to deform than ice at its pressure melting point. For example, under a given stress, ice at 0 °C deforms at a rate 100 times faster than ice at –20 °C (Hambrey, 1994). The presence of solid (nonvolatile) impurities also affects rheology, with debris–ice mixtures sharply increasing ice strength. Particulate concentrations less than 10% have a negligible effect, but they significantly strengthen ice (compared with pure ice) when they exceed 10%. Conversely, mixtures with rock particles above 85% were found to be weaker than pure ice, probably due to loss of cohesive strength (Benn and Evans, 1998, p. 149). The presence of solid impurities (tephra particles) is likely to be ubiquitous in glaciers in regions affected by volcanic activity. Their influence on ice rheology is relatively predictable, by increasing ice strength and thus strain response time.

Hydraulics

In temperate glaciers, water is transmitted only extremely slowly along intergranular veins and crystal boundaries in ice (Nye and Frank, 1973). On the rapid time scale of subglacial eruptions, unfractured glacier ice is effectively impermeable, unless crevasses are present. An upper crevasse layer will dominate water flow in glacier ice, and it will also drain any overlying firn (Fountain, 1989). Snow and firn layers are porous and, therefore, permeable, and will transmit flowing water (Paterson, 1994; Fountain and Walder, 1998). During a subglacial eruption, water is thought to rapidly fill a vault. Meltwater may overflow via an upper permeable layer of snow, firn, and/or crevassed ice (Smellie, 2000, 2001). In glaciers lacking a permeable upper layer, the surrounding ice will be floated when the water level reaches ~90% of the glacier thickness and the vault will drain subglacially, as a jökulhlaup (Björnsson, 1988). However, the presence of preexisting subglacial tunnels in the vent area ensures that many glaciers will be "leaky" during eruptions. Thus basal drainage may prevent the meltwater from floating the ice sheet, and a lake may be established, as occurred in the Gjálp eruption (Gudmundsson et al., 1997, 2004).

Modeling the hydraulics of a subglacial vault or lake that is overflowing via an upper permeable snow/firn layer is difficult because of the unique combination of processes involved. For glaciovolcanic systems, the only treatment available so far is by Smellie (2006). The results, which are summarized briefly here, suggest that overflowing is an unproven but feasible process that may be established relatively quickly. The permeability of firn and snow to water is low, typically 10^{-10} –10^{-5}, respectively (Colbeck and Anderson, 1982; Fountain, 1989). However, conditions are very different between "normal" (meteoric) *wetting* at ambient temperatures, and volcano-induced *flooding* with warm volcanic-heated water. Heat transfer between the (warm) liquid and solid causes phase changes that continuously alter the hydraulic properties of the medium (porosity and permeability will increase). Melting less dense firn also requires substantially less energy than is required to melt a corresponding volume of glacier ice, and the energy requirement reduces still further if that firn is dirty. Smellie (2006) identified a *wetting front* that advances ahead of a *zone of melting*. The melting zone is fully melted or disaggregated at its rear. Thus the width of the melting zone remains relatively short and the lateral pressure gradient (e.g., dP/dx in the equation for Darcy flow) is buffered at higher values than can occur in water percolating through a nonmelting particulate aquifer. The water volume flux per unit area is therefore correspondingly higher. Irregular pressure pulses caused by percussive effects of underwater explosions and associated tsunami may also enhance the gravity-driven meltwater flow and help to replenish warm water at the melting zone. A pre-eruption ice sheet surface that slopes away from the vent region is probably important for quickly establishing an overflowing system. Once the meltwater exits onto the surrounding glacier surface, much more rapid piped or channel flow is established.

Both overflowing and basal drainage are probably unstable over periods of days and weeks. An overflowing ice spillway will be rapidly cut down by thermal erosion, and subglacial tunnels will suffer from a combination of intermittent blockages caused by collapses of the volcanic edifice and shrinkage in proportion to the meltwater flux and meltwater temperature. Either meltwater escape route may dominate at different times in an advancing lava-fed delta.

Influence of Glaciers with Different Thermal Regimes on Terrestrial Subglacial Eruptions

Temperate Glaciers

Because temperate glaciers are at their pressure-melting point throughout, they therefore have the lowest energy requirements (per unit mass) for melting compared with cold (polar, subpolar) glaciers. Temperate glaciers are also wet-based and they typically have well-developed basal drainage systems, particularly tunnels (Fountain and Walder, 1998), which volcanic-generated meltwater will exploit during a subglacial eruption. Because the glacier ice is not frozen to its bed, it can be floated if water-filled glacial vaults develop sufficient hydraulic pressure, leading to sudden and dramatic jökulhlaups (Björnsson, 1988). Unambiguous evidence exists, both observational (Gudmundsson et al., 1997, 2004) and in the rock record (Smellie, 2001, 2006), for basal meltwater escape during subglacial eruptions. Conversely, the existence and importance of overflowing vaults is less well established and remains conjectural, although it is consistent with observations of historical eruptions and evidence interpreted from some lava-fed deltas (Smellie, 2001, 2006). Overflowing is an important process in the theoretical model for tuya eruptions proposed by Smellie (2000, 2001) as a method for establishing a stable lake surface and generation of horizontal passage zones in lava-fed deltas. However, thermal erosion of an ice spillway (the lake overflow point) may be rapid in temperate systems (cf. Raymond and Nolan, 2002) and it is thus unclear whether an overflowing glacial system can ever create a lake surface that has a stable elevation for a long period. So long as a lake is present, lava-fed deltas can probably develop in situations of either overflowing or basal discharge, but in either case they will show distinctive lithofacies evidence, comprising variations in passage zone elevations, for unstable meltwater discharge and a variable lake surface (Smellie, 2006). An important distinction is the relative rate of meltwater accumulation versus (basal) escape. In cases where basal discharge is slow, meltwater will accumulate more rapidly in the vault and will either (1) float the surrounding ice (in glaciers with a thin or no upper permeable layer) or (2) will overflow (glaciers with a thick upper permeable [firn/crevasse] layer). In the latter, the surrounding ice sheet will not be floated subsequently because thermal erosion will lower the elevation of the spillway, and the glacier is effectively pinned for as long as overflowing continues.

As discussed above, a natural limit of ~750–1000 m may exist for the maximum thickness of an ice sheet capable of sustaining a tuya eruption. This is a consequence of the model by Smellie (2000, 2001). Although the interpretation is untested, there are no natural terrestrial examples of tuyas known to the author that exceed 1000 m in height (Fig. 7). This might be an indication that many tuyas are height limited as a consequence of overflowing. Conversely, in glaciers dominated by high basal meltwater discharge, the water level may be maintained at a level below that required to float the glacier. In a very thick glacier (>1000 m), it may thus be possible to construct a much taller edifice than in a tuya dominated by overflowing, although the passage zone ("water level") elevation will be well below the ice sheet surface.

Being relatively warm, temperate ice will melt and deform by flow comparatively rapidly (cf. Gudmundsson et al., 2004). Glacial lakes will therefore be relatively wide and ice walls will collapse and recede by melting ahead of an advancing lava-fed delta. The ice recession will remove support from the volcano flanks, leading to frequent flank collapses and numerous slump scars in the preserved sequences, which, because of their well-defined stratification, will be most conspicuous in the early formed (pre–lava-fed delta) subaqueous successions in mafic tuff cones (tindars).

For eruptions of most felsic magmas and possibly for some mafic eruptions (Gudmundsson et al., 2004), the thermal and rheological constraints (lower magmatic temperatures, higher viscosities, conductive cooling) yield lower melting efficiencies (Hoskuldsson and Sparks, 1997). This is also true regardless of the effects of complex cooling histories of felsic glasses, which are capable of prolonging elevated temperatures (Wilding et al., 2004). As a consequence, less space is melted in the overlying ice relative to the volume of magma emplaced, yielding positive eruption pressures, lifting the ice cover and permitting any meltwater to escape continuously (Hoskuldsson and Sparks, 1997; Tuffen et al., 2001, 2002a, 2002b, 2007). This may be the most important reason why there are no known lava-fed deltas for felsic magmas.

Figure 8, A–E, illustrates the principal lithofacies, sequence architecture, and development of mafic and gas-rich felsic tuyas erupted in temperate ice. For effusive felsic eruptions, see below.

Polar Glaciers

Polar glaciers lack free water, are frozen to their base, and meltwater cannot escape subglacially. Glacial vaults will form but require significant additional energy expenditure initially (compared with temperate ice) to bring the ice up to melting point. The vault will always overflow; hence there will be no morphological evidence for meltwater-derived landforms, such as channels and eskers, nor traction current deposits, except in the off-glacier proglacial region. Overflowing will take place irrespective of any upper firn or crevasse layer. Because it is colder, and also if it is dirty, polar glacier ice will not so readily flow in toward the vault or lake. It is also less liable to collapse. The additional thermal requirements for melting will probably ensure that any vault/lake will be narrower. With less space, the vault/lake will rapidly become infilled by erupted magmatic products. Any lava-fed deltas will be short and subaerial lava may overrun onto the surrounding ice sheet surface. Thus, following ablation, extensive fields of chaotic lava rubble will surround polar glacier mafic tuyas (Fig. 8, F–I). There is no limit to the maximum height of any edifice, the height being determined solely by the volume of magma discharged and the thickness of the ice. The greater stability of the ice walls will lead to fewer flank collapses and slump scars in the volcanic sequences. Because of the slow rate of melt-back, if the ice buttresses edifices for longer periods they may have generally higher aspect ratios than edifices formed in temperate ice.

Products of gas-rich felsic eruptions in polar ice will be similar to those formed in temperate ice (cf. Fig. 8, D and E). By contrast, for felsic eruptions that are effusive because they are either gas-poor or in which vesiculation is hindered by thick ice overburden pressures, felsic domes and flow-dominated tuyas will be formed (Fig. 8, J–N). If a flow-dominated tuya continues activity after it penetrates the ice sheet surface (Fig. 8, L–N), lava will expand laterally, similar to extrusion of a tube of toothpaste. After ice sheet decay, any lava emplaced on the former ice sheet surface will collapse to form a field of lava rubble, similar to mafic tuyas but much less extensive.

Subpolar Glaciers

Subpolar glaciers are cold and thus resemble polar glaciers, but they differ significantly by having wet bases. Cold glaciers have the greatest thicknesses of firn (up to 125 m; Paterson, 1994) and crevasses are slower to deform and may extend to greater depths than in temperate ice. Two scenarios are likely for eruptions in subpolar glaciers.

(1) Because of the additional energy requirements for melting cold polar ice (e.g., ~16% more energy may be required to melt ice at −30 °C), the melting efficiency will be reduced relative to eruptions in temperate ice. Positive pressures will be generated for felsic eruptions and probably even for some mafic eruptions (cf. Gudmundsson et al., 2004). As the glacier is decoupled from its bed, it will be lifted and any meltwater will escape subglacially throughout the eruption. Thus no water-filled vault or lake will form, or lava-fed deltas, mafic tuya landforms, or a pillow lava pile. The edifices will likely be dominated by phreatomagmatic tephra, possibly capped by subaerial lava, or else may be lava flow dominated. Because of the buttressing effects of the enclosing slowly melting ice, any edifices will probably be relatively tall, as in polar glaciers.

(2) In many terrestrial situations, the temperature at depths greater than a few hundred meters in an ice sheet will probably be significantly warmer than −30 °C. Thus the additional energy required to melt the ice may not be particularly large. For felsic eruptions, which may contain just half of the thermal energy of mafic magmas (Hoskuldsson and Sparks, 1997), the greater energy demanded by the colder glacier will ensure that positive eruption pressures are still likely in all cases. For mafic magmas, which are erupted at higher temperatures, the thermal energy available and greater heat exchange efficiency may result in negative eruption pressures and enable a meltwater vault to form. Thereafter, the eruption will proceed to construct some combination of pillow and/or tephra mound and lava-fed delta(s), i.e., a mafic tuya, with complications related to overflowing and basally discharging meltwater as described for temperate glaciers. If overflowing occurs, the additional energy required to melt polar ice may ensure that the ice spillway is eroded more slowly compared to temperate ice situations. The ice sheet thickness limitation on heights of tuyas formed in overflowing systems, discussed above and with the same reservations, might also apply to eruptions in subpolar ice sheets. However, it is speculated that any lava-fed deltas are likely to be short and terminate in rubble fields; similar to those erupted in polar ice.

Implications of Mars's Environmental Conditions for Subglacial Volcanic Landforms

Although the present thin atmosphere and frigid climate of Mars does not allow liquid water to be stable in near-surface rocks or soil in equatorial areas, the planet is a significant repository for ice in polar caps and in the ground as pore fillings or segregated masses. The view from Viking showed flow lobes composed of crater ejecta, possible thermokarst features such as chaotically collapsed ground and irregular depressions, and numerous fluvial valleys and outflow channels that attest to a large source reservoir of ground ice, near the surface or at depth, in the planet's past (Carr, 1996). Evidence has also been interpreted in support of former frozen lakes and a putative paleo-ocean, also probably frozen, in the northern lowland plains and even the equatorial region, formerly much more extensive ice sheets, and ice caps on topographical highs (Hodges and Moore, 1979; Kargel and Strom, 1992; Parker et al., 1993; Chapman, 1994; Kargel et al., 1995; Helgason, 1999; Head and Pratt, 2001; Head and Marchant, 2003; Neukum et al., 2004; Murray et al., 2005). The Martian chasmata and outflow channels may also once have been ice filled (Lucchitta, 1982, 2001; Lucchitta et al., 1994; Chapman and Tanaka, 2001). Crater density data also suggest that volcanism has continued throughout Mars's geological history (Neukum et al., 2004) and probably interacted with this ancient store of ice. Numerous Mars landforms have been interpreted as volcanic edifices constructed within ice in each of these settings (e.g., Hodges and Moore, 1979; Chapman, 1994; Lucchitta et al., 1994; Chapman and Tanaka, 2001; Ghatan and Head, 2002; Chapman and Smellie, 2007). Melting associated with some of these edifices may have been the source of megascale outburst floods (e.g., Chapman et al., 2003).

Nonglacial environmental effects, such as Mars's lower gravity, lower atmospheric pressure, and thinner atmosphere, can have a significant effect on subaerial volcanic landforms (Wilson and Head, 1994; Head and Wilson, 2002). However, for subglacially erupted edifices, it is the presence and physical characteristics of the enclosing ice that will primarily determine the gross morphology and lithofacies, as on Earth. At present, assigning a subglacial origin to Mars's volcanoes is based almost wholly on morphology, and three fundamental candidate edifice types have been identified. In the following discussion relating to Mars's

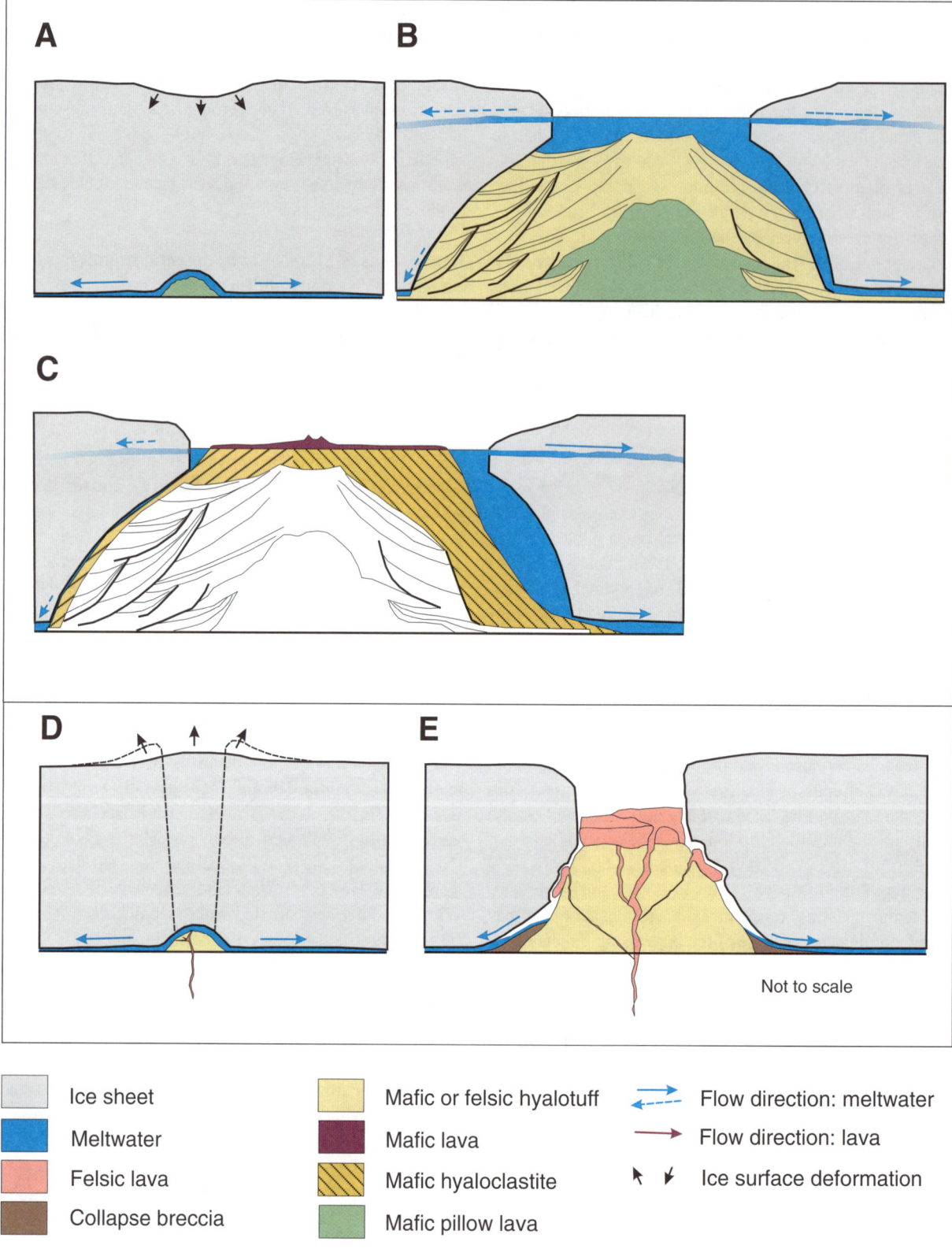

Figure 8 (*continued on next page*).

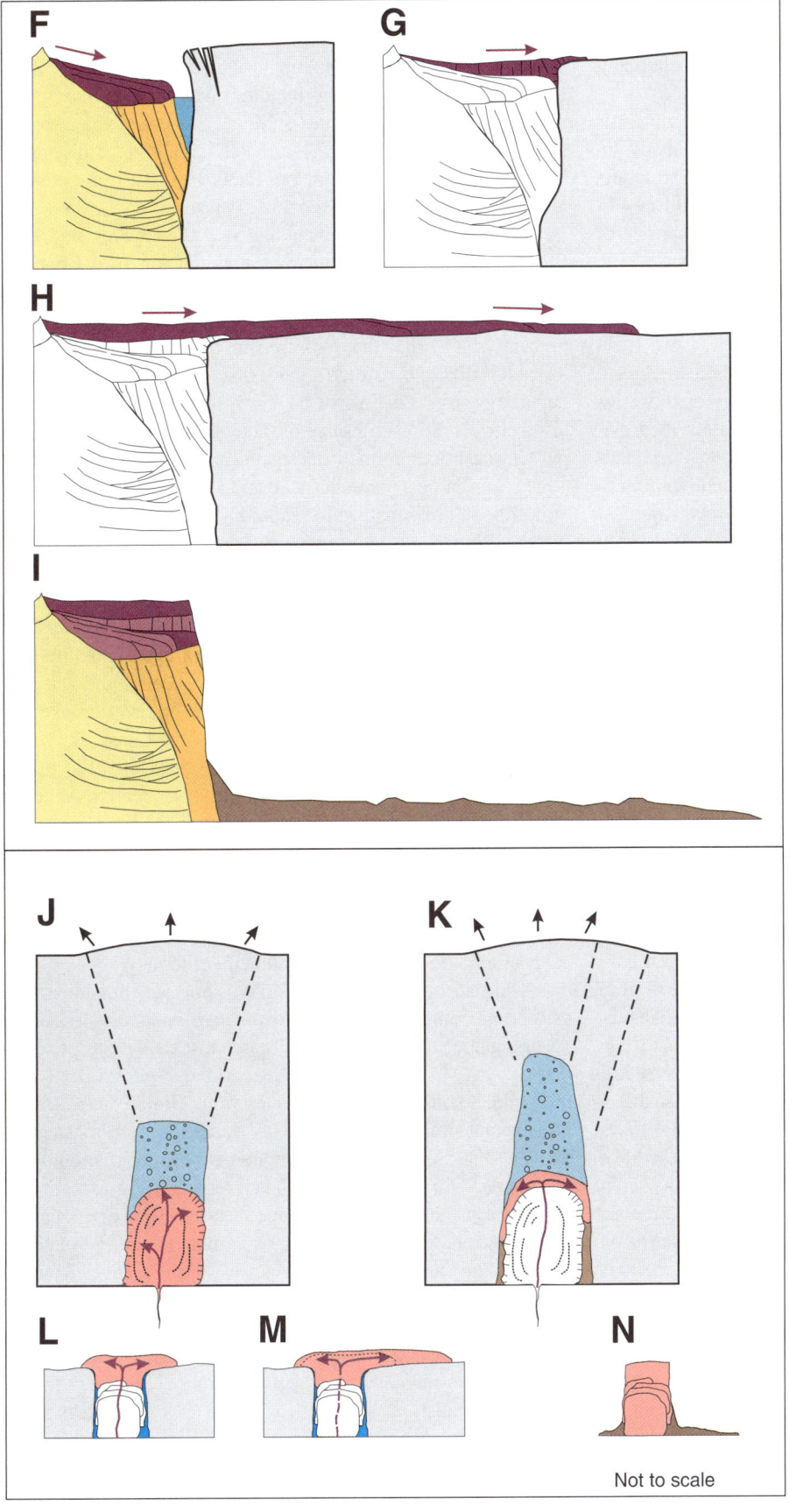

Figure 8 (*on this and previous page*). Cartoons showing stages of subglacial volcano formation under temperate (A–E) and polar (F–N) glaciers, for mafic and felsic tuyas. A–C—Mafic tuya. The pillow lava core shown is very common in these tuyas but may not always be present. Note the conspicuous slump scar surfaces (heavy lines) and stratification, some back-tilted, in the mafic subaqueous tephra pile, and meltwater escaping basally and by overflowing. Passage zone elevation variations in the lava-fed delta are not shown. D and E—Tephra-dominated felsic tuya. Explosive eruptions may penetrate the full glacier thickness from the earliest stages if the glacier is thin (dashed lines in D). Although wet, the vault is drained continuously during eruption. The tephra appear massive but may be faintly stratified. F–I—Mafic tuya. The glacier is frozen to its bed and meltwater escapes by overflowing (not shown). The narrow meltwater lake in F is rapidly infilled and the surrounding ice sheet surface overridden by subaerial lava (G), which collapses to form an extensive field of lava rubble following ice sheet decay (I). The associated tephra cone probably lacks the large numbers of conspicuous slump scar surfaces that are characteristic of eruptions in temperate glaciers (cf. B and C); note back-tilted tephra bedding. J–N—Felsic lava flow-dominated tuya. For effusive eruptions in thick ice (J and K), a tall cylindrical construct is formed. Meltwater will escape by overflowing (not shown). In eruptions through thinner ice (L–N), the construct may penetrate the ice sheet and expand laterally. Following ice sheet decay, a field of lava rubble will be formed (N) similar to that formed during mafic tuya eruptions in polar ice sheets, but less extensive. Effusive felsic eruptions under temperate ice will show similar features. Felsic domes and lobes may be the early formed parts of eruptions that ceased before they grew into tuyas. See text for further discussion.

conditions, two principal environmental assumptions are used: (i) any ice is unconfined surface ice, and (ii) it is water ice. The influence of ground ice (cryolithosphere), CO_2 ice, and clathrates is not considered.

The key to understanding any Mars glacial environment is knowledge of the potential ice thermal regime and density evolution, which will affect glacier hydrology, rheology, and structure. With its current very cold, hyperarid environment and low surface pressures, liquid water on Mars will rapidly evaporate or freeze, probably confining surface ice to polar caps (Carr, 1979). Conversely, former large unconfined surface-ice sheets may have existed at widely differing latitudes in previous periods of Mars's history (e.g., Kargel and Strom, 1992; Kargel et al., 1995; Head and Pratt, 2001; Ghatan and Head, 2002; Head and Marchant, 2003; Neukum et al., 2004). The thermal regime of those ice sheets is uncertain. Head and Marchant (2003) described evidence for a former cold-based ("polar") glacier in equatorial latitudes. Others have cited abundant features consistent with meltwater processes involved in the collapse of ice sheets (Kargel et al., 1995; Ghatan and Head, 2002). The hypothesized wet-based ice sheets may not have been fully temperate but may have been polythermal (subpolar), with basal melting conditions caused by enhanced geothermal heat-flow gradients. Thus, across the broad expanse of Mars's time and geography, the thermal regime of unconfined surface ice sheets is not well constrained; it may have been temperate, polar, and/or subpolar at different times and/or in different regions.

Aspect ratios for selected freestanding candidate subglacially formed volcanic edifices on Mars are plotted in Figure 9. Supposed analogous volcanic edifices in the chasmata (i.e., interior layered deposits [ILDs]; Chapman and Tanaka, 2001) are excluded; the shapes of ILDs are highly irregular and the walls of the chasmata confine the ILDs in part. The selected landforms comprise three basic morphological types: low domes, tall cones, and flat-topped, steep-sided landforms. Data for terrestrial constructs are also plotted. In Figure 9A, dimensions of terrestrial felsic domes, pillow mounds/sheets, and subglacial sheetlike outcrops are substantially lower and less extensive than Mars's low domes. Under the low Mars gravity and buttressing effects of ice, any Mars felsic domes (should felsic compositions exist there) should be slim and taller than on Earth. The confining effects and thickness of the ice will principally determine the edifice height. Even if the ice is completely melted through, however, it is hard to envisage how the rheology and thermal energy of felsic domes would enable them to extend laterally *for a few tens of km* through ice sheets of any thermal regime to reach the dimensions of the Martian examples. For example, in a voluminous effusive eruption that penetrated an ice sheet, felsic magma would spread out over the surrounding ice sheet surface. Following ablation, any lava emplaced on the ice would collapse to form a layer of lava rubble surrounding a tall, slim in situ "core" (cf. Fig. 8, L–N). Conversely, dimensional data for terrestrial mafic sheetlike outcrops of Dalsheidi type suggest that they have the ability to extend laterally for large distances while remaining relatively thin. They were probably emplaced as sills injected and inflated along the bedrock–ice interface (Wilson and Head, 2002). Mars's low domes might thus be exhumed subglacially intruded "interface sills" inflated to a few hundred meters in thickness. The much greater magma volumes of Mars's low domes are also consistent with predictions suggesting that large magma discharges are possible on Mars (Wilson and Head, 1994). It is not possible to infer a thermal regime from those outcrops.

Tall cones on Mars, including ridge landforms in Utopia Planitia that are conical in transverse section (see Chapman [1994, fig. 6] and Ghatan and Head [2002] for context), appear to be dispersed about a relatively constant aspect ratio, which is significantly *lower* than that for terrestrial tephra mounds (Fig. 9B). Volcanic edifices constructed in ice of any thermal regime should have large aspect ratios, as on Earth (Fig. 7). The very much lower aspect ratios on Mars are more akin to those expected in pyroclastic cones formed during unconfined subaerial or subaqueous eruptions, particularly under the lower Mars gravity and atmospheric pressures, which should lead to greater ash particle dispersal by fallout and pyroclastic density flows compared to Earth (Wilson and Head, 1994; Head and Wilson, 2002). Chapman (1994, 2003) used the presence and inferred similarities of ridges in Utopia Planitia to Icelandic tindars as support for a former lake. Although Chapman (1994, 2003) assumed the lake was frozen, the alternative, that it may have been liquid when the ridges formed, should also be considered, or else the lake was absent. Moreover, from their aspect ratios and profiles, Mars's tall cones are almost certainly much too thick (> 1000 m) for them to be far-traveled subglacial sheetlike sequences. Although the Utopia Planitia examples overlap with terrestrial sheetlike outcrops, the morphology of all of the Mars examples, as cones and conical ridges, is entirely unlike any of the terrestrial occurrences so far described. Therefore it is difficult to relate the tall cones on Mars to a glacial eruptive environment, at least on the basis of admittedly sparse morphometry criteria.

The flat-topped Mars landforms, some with broad summit craters, are amongst the most distinctive constructs on Mars. At first glance, they might seem to be most easily interpreted as mafic tuyas (Chapman, 1994; Ghatan and Head, 2002; Chapman and Smellie, 2007). Examples from Utopia Planitia have dimensions similar to terrestrial tuyas, and they are most easily explained as such (Fig. 9C), although we cannot rule out other mechanisms for their formation. Their relatively low aspect ratios and flat tops, suggesting extensive lava-fed delta progradation, are characteristic of tuya eruptions emplaced within temperate ice. This may also be true for the lowest of the edifices from the Dorsa Argentea Formation, with a height comparable to several terrestrial mafic tuyas but containing a substantially greater volume; it may thus be a mafic tuya that underwent a phase of very extensive lava-fed delta progradation during a voluminous eruption. By contrast, all of the other candidate tuya edifices in the Dorsa Argentea Formation are substantially taller and wider than mafic tuyas on Earth. A plausible explanation is that they were erupted into ice sheet(s) ~2 km thick, followed by lateral delta expansion (Ghatan and Head, 2002). Such extensive delta progradation is only likely to occur within easily melted temperate ice, consistent with the observations of possible meltwater effects (channels, eskers) observed in the same area by Ghatan and Head (2002). The edifice heights of those Mars edifices considerably exceed the ~ 750–1000 m height limit on mafic tuyas constructed in overflowing terrestrial systems. There should be no change to this hypothetical limit under Mars conditions. Thus the Mars examples might have been emplaced in a wet-based (temperate) ice sheet that was dominated by basal meltwater discharge. By contrast, mafic tuyas formed in polar or subpolar ice sheets should have similar aspect ratios to the main group of terrestrial tuyas (group a in Fig. 7) or extend to somewhat *higher* ratios (cf. Fig. 8, F–I). They should also be surrounded by laterally extensive fields of rubble (the remains of lavas that flowed over the ice sheet surface), which have not yet been identified.

Finally, Chapman and Tanaka (2001) identified several large landforms (interior layered deposits [ILDs]) resembling lava-fed deltas within some Valles Marineris chasmata. The existence and origin of ice in the chasmata are uncertain but Carr (1979) suggested that the water might have been sourced in seepage from confined aquifers cut across tectonically by the chasmata walls. This should yield an ice sheet formed wholly of ice, i.e., lacking any upper firn layer, although crevasses may be present in regions undergoing extension (e.g., crossing basement riegels). The deltalike landforms are up to nearly 8 km thick, which would make them the thickest lava-fed deltas in the Solar System. Some are laterally extensive (up to 100 km) and they are not surrounded by lava rubble fields. On both counts, eruption and emplacement in cold polar ice would seem to be discounted. Emplacement of the chasmata ILD landforms through wet-based ice with dominant basal drainage might be more plausible based on the present analysis, and is supported by observations of possible glacial erosion in Candor Chasma (Chapman et al., 2004). However, there are no descriptions of any associated meltwater-influenced landforms on the chasmata floors. Conversely, Chapman (1994, 2003) suggested that any Valles Marineris ice might be polar, assuming a Mars climate similar to the present. Chapman and Smellie (2007) also noted the restriction of outflow channels to the chasmata mouths, as if cut by periglacial runoff sourced on the surface of polar glaciers (see also Chapman et al., 2003). Evidence for the thermal regime of any glacial ice within the Mars chasmata is thus ambiguous and conflicting and cannot yet be resolved.

From our assessment of the comparative morphometry of three principal candidate subglacial volcanic landforms on Mars, there is least difficulty in assigning a glaciovolcanic origin to the Mars flat-topped landforms. However, as this comparison is based solely on morphometry, which is a coarse-resolution criterion as used here, we cannot rule out other origins until we are able to conduct much more detailed studies, ideally with stratigraphic details of the sequences present (e.g., Chapman and Tanaka, 2001). The flat-topped landforms on Mars are consistently identified as plausible mafic tuyas, although most are also products of substantially

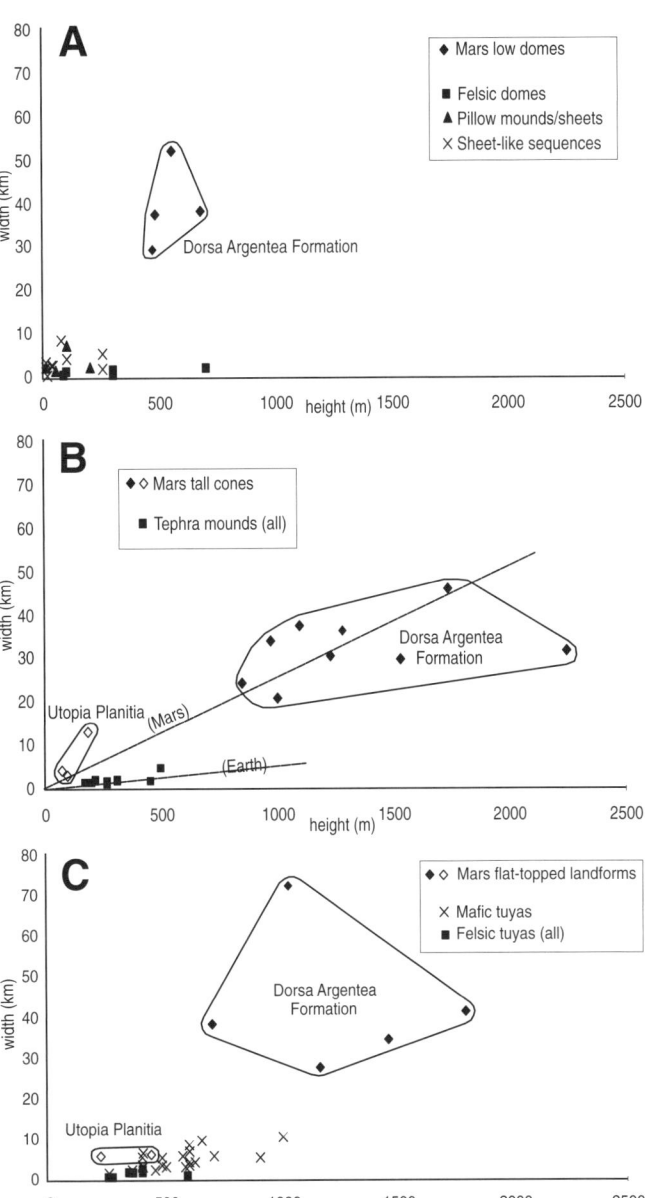

Figure 9. Diagrams showing aspect ratios (height vs. width) for morphologically distinctive Mars landforms, which may have been erupted subglacially. Also shown in each diagram are the terrestrial landform types intuitively most likely to be similar to the Mars examples. See text for discussion. Mars data from Chapman (1994) and Ghatan and Head (2002). Data for Mars ILDs not plotted.

greater magma discharges than have been observed in terrestrial analogs. The exceptionally large volumes of the edifices (compared to Earth) would also perhaps be easier to explain if they are polygenetic volcanoes, formed over longer periods of time. Thus the Dorsa Argentea Formation edifices may be compound, formed by the products of multiple eruptions, possibly as stacked lava-fed deltas constructed through a much thinner but continuously re-forming ice sheet. A plausible terrestrial analogy for this type of edifice is the long-lived (6–10 m.y.) Neogene James Ross Island stratovolcano, Antarctica, which is formed of multiple superimposed mainly glacially emplaced lava-fed deltas (Nelson, 1975; Smellie, 2006; Smellie et al., 2008). Alternatively, eruptions from multiple overlapping vents at different times within a long-lived much thicker ice sheet could also construct an areally large edifice. The final stage of eruptions might then penetrate the ice sheet surface and spread laterally as extensive lava-fed deltas, thus forming the "flat" summit region. Mars's low domes have no obvious terrestrial analogs but they are speculatively compared with far-traveled and magmatically inflated subglacial "interface sills" exhumed following ice sheet ablation. By contrast, Mars's tall cones and conical ridges seem to have aspect ratios too low to be likely subglacial tephra mounds. They are more easily interpreted as pyroclastic cones formed during subaerial or subaqueous eruptions; i.e., under nonglacial conditions.

CONCLUSIONS

The interpretation of subglacial volcanic landforms on Earth does not only indicate the presence or absence of ice sheets. The features can yield a more holistic range of paleoenvironmental parameters, including: ice thickness, thermal regime, and surface elevation. At least nine types of terrestrial monogenetic subglacial volcanic successions are identified, based on morphology, morphometry, lithofacies, and sequence architecture. Each of these properties is influenced variably by magma composition (mafic, felsic), the simple presence and thickness of glacial ice, and three principal glacier parameters (thermal regime, rheology, and hydraulics), thus providing a possible way of assessing eruptive paleoenvironments for candidate volcanic landforms on Mars. Only three morphological types of candidate landforms have been described for Mars. Using analogies for terrestrial subglacial eruptions, interpretations of some of the landforms on Mars as putative subglacial volcanic edifices are confirmed although all have significantly larger volumes than possible terrestrial analogs. Mars's low domes might be subglacial sheetlike sequences, although the glacial thermal regime is unspecific. Mars's flat-topped edifices consistently resemble mafic tuya-like edifices emplaced within temperate ice up to 2 km thick. The flat-topped constructs are far larger than any known on Earth, which may be an indication of very long-lived magmatism, much higher magma discharge, and/or polygenetic edifices. If polygenetic, the Mars landforms might have formed either by (i) multiple sub-ice eruptions within a very thick ice sheet that terminated in extensive lava-fed delta(s), or (ii) multiple lava-fed delta eruptions beneath a much thinner ice sheet, which remained thin during the eruptive period but re-formed repeatedly on the volcano after each eruptive phase. A plausible terrestrial analogy for the latter mode of volcano construction is the long-lived James Ross Island stratovolcano in Antarctica. The origin of Mars's tall cones is still enigmatic. Their aspect ratios most resemble those of subaerial or subaqueous pyroclastic cones formed by eruptions in a nonglacial (lake/sea) setting. However, all of the comparisons in this paper are based solely on coarse-resolution morphometry. Other origins cannot be ruled out until much more detailed studies are conducted, ideally with stratigraphic details of the sequences present.

ACKNOWLEDGMENTS

I am grateful to numerous people with whom I have conducted a dialog over many years, but I particularly wish to thank the following who were very helpful in my preparation for this paper: Melanie Kelman, Gudrún Larsen, Kristján Sæmundsson, and Sue Loughlin (edifice/sequence dimensions); Rob Arthern, Chris Doake, Richard Hindmarsh, and David Vaughan (glacier physics); Mary Chapman (Mars volcanic edifices); and Dave McGarvie, Hugh Tuffen, and John Stevenson (Icelandic subglacial felsic systems). I am also grateful for the positive review on this paper by W. L. Jaeger.

REFERENCES CITED

Benn, D.I., and Evans, D.J.A., 1998, Glaciers and glaciation: London, Arnold, 734 p.

Bergh, S.G., and Sigvaldason, G.E., 1991, Pleistocene mass-flow deposits of basaltic hyaloclastite on a shallow submarine shelf, South Iceland: Bulletin of Volcanology, v. 53, p. 597–611, doi: 10.1007/BF00493688.

Björnsson, K., 1988, Hydrology of ice caps in volcanic regions: Vísindafélag Íslendinga, Societas Scientarium Islandica, v. 45, 139 p.

Carr, M.H., 1979, Formation of Martian flood features by release of water from confined aquifers: Journal of Geophysical Research, v. 84, p. 2995–3007, doi: 10.1029/JB084iB06p02995.

Carr, M.H., 1996, Water on Mars: Oxford, UK, Oxford University Press, 229 p.

Carsewell, D.A., 1983, The volcanic rocks of the Solheimajökull area, southern Iceland: Jökull, v. 33, p. 61–71.

Chapman, M.G., 1994, Evidence, age, and thickness of a frozen paleolake in Utopia Planitia, Mars: Icarus, v. 109, p. 393–406, doi: 10.1006/icar.1994.1102.

Chapman, M.G., 2003, Sub-ice volcanoes and ancient oceans/lakes: A Martian challenge: Global and Planetary Change, v. 35, p. 185–198, doi: 10.1016/S0921-8181(02)00126-1.

Chapman, M.G., and Tanaka, K.L., 2001, The interior trough deposits on Mars: Sub-ice volcanoes?: Journal of Geophysical Research, v. 106, p. 10,087–10,100, doi: 10.1029/2000JE001303.

Chapman, M.G., Gudmundsson, M.T., Russell, A.J., and Hare, T.M., 2003, Possible Juventae Chasma subice volcanic eruptions and Maja Valles ice outburst floods on Mars: Implications of Mars Global Surveyor crater densities, geomorphology, and topography: Journal of Geophysical Research, v. 108, doi: 10.1029/2002JE002009.

Chapman, M.G., and Smellie, J.L., 2007, Mars interior layered deposits and terrestrial sub-ice volcanoes compared: Observations and interpretations of similar geomorphic characteristics, in Chapman, M.G., ed., The geology of Mars: Evidence from Earth-based analogs: Cambridge, UK, Cambridge University Press, p. 178–210.

Chapman, M.G., Soderblum, L.A., and Cushing, G., 2004, Evidence for very young glacial processes in central Candor Chasma, Mars: Lunar and Planetary Institute, March 15–19, Houston, Texas, LPSC 36th CD, #1850.

Colbeck, S.C., and Anderson, E.A., 1982, The permeability of a melting snow cover: Water Resources Research, v. 18, p. 904–908, doi: 10.1029/WR018i004p00904.

Dixon, J.E., Filiberto, J.R., Moore, J.G., and Hickson, C.J., 2002, Volatiles in basaltic glasses from a subglacial volcano in northern British Columbia (Canada): Implications for ice sheet thickness and mantle volatiles, in Smellie, J.L., and Chapman, M.G., eds., Volcano—ice interaction on Earth and Mars: London, Geological Society, Special Publication, v. 202, p. 255–271.

Edwards, B.R., and Russell, J.K., 2002, Glacial influences on morphology and eruptive products of Hoodoo Mountain volcano, Canada, in Smellie, J.L., and Chapman, M.G., eds., Volcano—ice interaction on Earth and Mars: London, Geological Society, Special Publication, v. 202, p. 179–194.

Fountain, A.G., 1989, The storage of water in, and hydraulic characteristics of, the firn of South Cascade Glacier, Washington State, U.S: Annals of Glaciology, v. 13, p. 69–75.

Fountain, A.G., and Walder, J.S., 1998, Water flow through temperate glaciers: Reviews of Geophysics, v. 36, p. 299–328, doi: 10.1029/97RG03579.

Furnes, H., and Fridliefsson, I.B., 1979, Pillow block breccia—Occurrences and mode of formation: Neues Jahrbuch für Geologie und Paläontologie, v. 3, p. 147–154.

Ghatan, G.J., and Head, J.W., 2002, Candidate subglacial volcanoes in the south polar region of Mars: Morphology, morphometry, and eruption condition: Journal of Geophysical Research, v. 107, p. 5048, doi: 10.1029/2001JE001519.

Gudmundsson, M.T., 2003, Melting of ice by magma-ice-water interactions during subglacial eruptions as an indicator of heat transfer in subaqueous eruptions, in White, J.D.L., Smellie, J.L., and Clague, D.A., eds., Subaqueous explosive volcanism: American Geophysical Union, Geophysical Monograph, v. 140, p. 61–72.

Gudmundsson, M.T., Pálsson, F., Björnsson, H., and Högnadóttir, T., 2002, The hyaloclastite ridge formed in the subglacial 1996 eruption of Gjálp, Vatnajökull, Iceland: Present day shape and future preservation, in Smellie, J.L., and Chapman, M.G., eds., Volcano—ice interaction on Earth and Mars: London, Geological Society, Special Publication, v. 202, p. 319–335.

Gudmundsson, M.T., Sigmundsson, F., and Björnsson, H., 1997, Ice-volcano interaction of the 1996 Gjálp subglacial eruption, Vatnajökull, Iceland: Nature, v. 389, p. 954–957, doi: 10.1038/40122.

Gudmundsson, M.T., Sigmundsson, F., Björnsson, H., and Högnadóttir, T., 2004, The 1996 eruption at Gjálp, Vatnajökull ice cap, Iceland: Efficiency of heat transfer, ice deformation and subglacial water pressure: Bulletin of Volcanology, v. 66, p. 46–65, doi: 10.1007/s00445-003-0295-9.

Hambrey, M.J., 1994, Glacial environments: London, UCL Press Ltd., 296 p.

Head, J.W., and Pratt, S., 2001, Extensive Hesperian-aged south polar ice sheet on Mars: Evidence for massive melting and retreat, and lateral flow and ponding of meltwater: Journal of Geophysical Research, v. 106, p. 12,275–12,299, doi: 10.1029/2000JE001359.

Head, J.W., and Wilson, L., 2002, Mars: A review and synthesis of general environments and geological settings of magma—H_2O interactions, in Smellie, J.L., and Chapman, M.G., eds., Volcano—ice interaction on Earth and Mars: London, Geological Society, Special Publication, v. 202, p. 27–57.

Head, J.W., and Marchant, D.R., 2003, Cold-based mountain glaciers on Mars: Western Arsia Mons: Geology, v. 31, p. 641–644, doi: 10.1130/0091-7613(2003)031<0641:CMGOMW>2.0.CO;2.

Helgason, J., 1999, Formation of Olympus Mons and the aureole-escarpment problem on Mars: Geology, v. 27, p. 231–234, doi: 10.1130/0091-7613(1999)027<0231:FOOMAT>2.3.CO;2.

Hickson, C.J., 2000, Physical controls and resulting morphological forms of Quaternary ice-contact volcanoes in western Canada: Geomorphology, v. 32, p. 239–261, doi: 10.1016/S0169-555X(99)00099-9.

Hodges, C.A., and Moore, H.J., 1979, The subglacial birth of Olympus Mons and its aureoles: Journal of Geophysical Research, v. 84, p. 8061–8074, doi: 10.1029/JB084iB14p08061.

Honnorez, J., and Kirst, P., 1975, Submarine basaltic volcanism: Morphometric parameters for discriminating hyaloclastites from hyalotuffs: Bulletin of Volcanology, v. 39, p. 1–25.

Höskuldsson, A., and Sparks, R.S.J., 1997, Thermodynamics and fluid dynamics of effusive subglacial eruptions: Bulletin of Volcanology, v. 59, p. 219–230, doi: 10.1007/s004450050187.

Jones, J.G., 1969, Intraglacial volcanoes of the Laugarvatn region, south-west Iceland—I: Quarterly Journal of the Geological Society of London, v. 124, p. 197–211.

Jones, J.G., 1970, Intraglacial volcanoes of the Laugarvatn region, southwest Iceland: The Journal of Geology, v. 78, p. 127–140.

Kargel, J.S., and Strom, R.G., 1992, Ancient glaciation on Mars: Geology, v. 20, p. 3–7, doi: 10.1130/0091-7613(1992)020<0003:AGOM>2.3.CO;2.

Kargel, J.S., Baker, V.R., Begét, J.E., Lockwood, J.F., Péwé, T.L., Shaw, J.S., and Strom, R.G., 1995, Evidence of ancient continental glaciation in the Martian northern plains: Journal of Geophysical Research, v. 100, p. 5351–5368, doi: 10.1029/94JE02447.

Kelman, M.C., Russell, J.K., and Hickson, C.J., 2002, Effusive intermediate glaciovolcanism in the Garibaldi Volcanic Belt, southwestern British Columbia, Canada, in Smellie, J.L., and Chapman, M.G., eds., Volcano—ice interaction on Earth and Mars: London, Geological Society, Special Publication, v. 202, p. 195–211.

Larsen, G., 2002, Holocene eruptions within the Katla volcanic system, south Iceland: Characteristics and environmental impact: Jökull, v. 49, p. 1–28.

LeMasurier, W.E., 2002, Architecture and evolution of hydrovolcanic deltas in Marie Byrd Land, Antarctica, in Smellie, J.L., and Chapman, M.G., eds., Volcano—ice interaction on Earth and Mars: London, Geological Society, Special Publication, v. 202, p. 115–148.

Lescinsky, D.T., and Fink, J.H., 2000, Lava and ice interaction at stratovolcanoes: Use of characteristic faetures to determine past glacial extents and future volcanic hazards: Journal of Geophysical Research, v. 105, p. 23,711–23,726, doi: 10.1029/2000JB900214.

Loughlin, S.C., 2002, Facies analysis of proximal subglacial and proglacial volcaniclastic successions at the Eyjafjallajökull central volcano, southern Iceland, in Smellie, J.L., and Chapman, M.G., eds., Volcano—ice interaction on Earth and Mars: London, Geological Society, Special Publication, v. 202, p. 149–178.

Lucchitta, B.K., 1982, Ice sculpture in the Martian outflow channels: Journal of Geophysical Research, v. 87, p. 9951–9973, doi: 10.1029/JB087iB12p09951.

Lucchitta, B.K., 2001, Antarctic ice streams and outflow channels on Mars: Geophysical Research Letters, v. 28, p. 403–406, doi: 10.1029/2000GL011924.

Lucchitta, B.K., Isbell, N.K., and Howington-Kraus, A., 1994, Topography of Valles Marineris: Implications for erosional and structural history: Journal of Geophysical Research, v. 99, p. 3783–3798, doi: 10.1029/93JE03095.

Mathews, W.H., 1947, Tuyas, flat-topped volcanoes in northern British Columbia: American Journal of Science, v. 245, p. 560–570.

Mathews, W.H., 1951, The Table, a flat-topped volcano in southern British Columbia: American Journal of Science, v. 249, p. 830–841.

Mathews, W.H., 1958, Geology of the Mount Garibaldi area, southwestern British Columbia. Part II. Geomorphology and Quaternary volcanic rocks: Geological Society of America Bulletin, v. 69, p. 161–178, doi: 10.1130/0016-7606(1958)69[161:GOTMGM]2.0.CO;2.

Mee, K., Tuffen, H., and Gilbert, J.S., 2006, Snow-contact volcanic facies and their use in determining past eruptive environments at Nevados de Chillán volcano, Chile: Bulletin of Volcanology, v. 68, p. 363–376, doi: 10.1007/s00445-005-0017-6.

Murray, J.B., Muller, J.-P., Neukum, G., Werner, S.C., van Gasselt, S., Hauber, E., Markiewicz, W.J., Head, J.W., Foing, B.H., Page, D., Mitchell, K.L., Portyankina, G., and the HRSC Co-Investigator Team, 2005, Evidence from the Mars Express High Resolution Stereo Camera for a frozen sea close to Mars' equator: Nature, v. 434, p. 352–356, doi: 10.1038/nature03379.

Nelson, P.H.H., 1975, The James Ross Island Volcanic Group of north-east Graham Land: British Antarctic Survey Scientific Reports, v. 54, 62 p.

Neukum, G., Jaumann, R., Hoffmann, H., Hauber, E., Head, J.W., Basilevsky, A.T., Ivanov, B.A., Werner, S.C., van Gasselt, S., Murray, J.B., McCord, T., and the HRSC Co-Investigator Team, 2004, Recent and episodic volcanic and glacial activity on Mars revealed by the High Resolution Stereo Camera: Nature, v. 432, p. 971–979, doi: 10.1038/nature03231.

Nye, J.F., and Frank, F.C., 1973, Hydrology of the intergranular veins in a temperate glacier: Publication de l'Association Internationale d'Hydrologie Scientifique, v. 95, p. 157–161.

Parker, T.J., Gorsline, D.S., Saunders, R.S., Pieri, D.C., and Schneeberger, D.M., 1993, Coastal geomorphology of the Martian northern plains: Journal of Geophysical Research, v. 98, p. 11,061–11,078, doi: 10.1029/93JE00618.

Paterson, W.S.B., 1994, The physics of glaciers: Oxford, UK, Pergamon, 480 p.

Raymond, C.F., and Nolan, M., 2002, Drainage of a glacial lake through an ice spillway: IAHS Publ. v. 264, p. 199–207.

Skilling, I.P., 1994, Evolution of an englacial volcano: Brown Bluff, Antarctica: Bulletin of Volcanology, v. 56, p. 573–591, doi: 10.1007/BF00302837.

Skilling, I.P., 2002, Basaltic pahoehoe lava-fed deltas: Large-scale characteristics, clast generation, emplacement processes and environmental discrimination, in Smellie, J.L., and Chapman, M.G., eds., Volcano—ice interaction on Earth and Mars: London, Geological Society, Special Publication, v. 202, p. 91–113.

Smellie, J.L., 1999, Lithostratigraphy of Miocene—Recent, alkaline volcanic fields in the Antarctic Peninsula and eastern Ellsworth Land: Antarctic Science, v. 11, p. 362–378, doi: 10.1017/S0954102099000450.

Smellie, J.L., 2000, Subglacial eruptions, in Sigurdsson, H., ed., Encyclopaedia of Volcanoes, San Diego, Academic Press, p. 403–418.

Smellie, J.L., 2001, Lithofacies architecture and construction of volcanoes in englacial lakes: Icefall Nunatak, Mount Murphy, eastern Marie Byrd Land, Antarctica, in White, J.D.L., and Riggs, N.R., eds., Volcaniclastic sedimentation in lacustrine settings: IAS Special Publication, v. 30, p. 73–98.

Smellie, J.L., 2002, The 1969 subglacial eruption on Deception Island (Antarctica): Events and processes during an eruption beneath a thin glacier and implications for volcanic hazards, in Smellie, J.L., and Chapman, M.G., eds., Volcano-ice interaction on Earth and Mars: London, Geological Society Special Publication, v. 202, p. 59–79.

Smellie, J.L., 2006, The relative importance of supraglacial versus subglacial meltwater escape in basaltic subglacial tuya eruptions: An important unresolved conundrum: Earth-Science Reviews, v. 74, p. 241–268, doi: 10.1016/j.earscirev.2005.09.004.

Smellie, J.L., 2008, Basaltic subglacial sheet-like sequences: Evidence for two types with different implications for the inferred thickness of associated ice: Earth-Science Reviews, v. 88, p. 60–88.

Smellie, J.L., and Skilling, I.P., 1994, Products of subglacial eruptions under different ice thicknesses: Two examples from Antarctica: Sedimentary Geology, v. 91, p. 115–129.

Smellie, J.L., and Hole, M.J., 1997, Products and processes in Pliocene-Recent, subaqueous to emergent volcanism in the Antarctic Peninsula: Examples of englacial Surtseyan volcano construction: Bulletin of Volcanology, v. 58, p. 628–646, doi: 10.1007/s004450050167.

Smellie, J.L., and Chapman, M.G., eds., 2002, Volcano—ice interaction on Earth and Mars: London, Geological Society, Special Publication, v. 202, 431 p.

Smellie, J.L., Hole, M.J., and Nell, P.A.R., 1993, Late Miocene valley-confined subglacial volcanism in northern Alexander Island, Antarctic Peninsula: Bulletin of Volcanology, v. 55, p. 273–288, doi: 10.1007/BF00624355.

Smellie, J.L., McIntosh, W.C., and Esser, R., 2006, Eruptive environment of volcanism on Brabant Island: Evidence for thin wet-based ice in northern Antarctic Peninsula during the late Quaternary: Palaeogeography, Palaeoclimatology, Palaeoecology, v. 231, p. 233–252, doi: 10.1016/j.palaeo.2005.07.035.

Smellie, J.L., Johnson, J.S., McIntosh, W.C., Esser, R., Gudmundsson, M.T., Hambrey, M.J., and van Wyk de Vries, B., 2008, Six million years of glacial history recorded in the James Ross Island Volcanic Group, Antarctic Peninsula: Palaeogeography, Palaeoclimatology, Palaeoecology, v. 260, p. 142–148.

Snorrason, S.P., and Vilmundardóttir, E.G., 2000, Pillow lava sheets: Origins and flow patterns, in Gulick, V.C., and Gudmundsson, M.T., eds., Volcano/ice interaction on Earth and Mars, Conference Abstracts, Reykjavik, Iceland, August 13–18, 62 p. (http://astrogeology.usgs.gov/Projects/VolcanoIceWorkshop/).

Staudigel, H., and Schmincke, H.-U., 1984, The Pliocene seamount series of La Palma/Canary Islands: Journal of Geophysical Research, v. 89, p. 11,195–11,215, doi: 10.1029/JB089iB13p11195.

Stevenson, J., McGarvie, D.W., Smellie, J.L., and Gilbert, J.S., 2006, Subglacial and ice-contact volcanism at the Öræfajökull stratovolcano, Iceland: Bulletin of Volcanology, v. 68, p. 737–752, doi: 10.1007/s00445-005-0047-0.

Stevenson, J.A., Smellie, J.L., McGarvie, D.W., and Gilbert, J.S., 2009, Subglacial intermediate volcanism at Kerlingarfjoll, Iceland: Magma-water interaction beneath thick ice: Journal of Volcanology and Geothermal Research (in press).

Tuffen, H., Gilbert, J., and McGarvie, D., 2001, Products of an effusive subglacial rhyolite eruption: Bláhnúkur, Torfajökull, Iceland: Bulletin of Volcanology, v. 63, p. 179–190, doi: 10.1007/s004450100134.

Tuffen, H., McGarvie, D.W., Gilbert, J.S., and Pinkerton, H., 2002a, Physical volcanology of a subglacial-to-emergent rhyolite tuya at Rauđufossafjöll, Torfajökull, Iceland, in Smellie, J.L., and Chapman, M.G., eds., Volcano—ice interaction on Earth and Mars: Geological Society of London, Special Publication, v. 202, p. 213–236.

Tuffen, H., Pinkerton, H., McGarvie, D.W., and Gilbert, J.S., 2002b, Melting of the glacier base during a small-volume subglacial rhyolite eruption: Evidence from Bláhnúkur, Iceland: Sedimentary Geology, v. 149, p. 183–198, doi: 10.1016/S0037-0738(01)00251-2.

Tuffen, H., McGarvie, D.W., and Gilbert, J.S., 2007, Will subglacial rhyolite eruptions be explosive or intrusive? Some insights from analytical models: Annals of Glaciology, v. 45, p. 87–94.

van Bemmelen, R.W., and Rutten, M.G., 1955, Table mountains of northern Iceland: Leiden, Brill, 217 p.

Walker, G.P.L., and Blake, D.H., 1966, The formation of a palagonite breccia mass beneath a valley glacier in Iceland: Quarterly Journal of the Geological Society of London, v. 122, p. 45–61.

Werner, R., and Schmincke, H.-U., 1999, Englacial vs lacustrine origin of volcanic table mountains: Evidence from Iceland: Bulletin of Volcanology, v. 60, p. 335–354, doi: 10.1007/s004450050237.

White, J.D.L., Smellie, J.L., and Clague, D.A., 2003, A deductive outline and topical overview of subaqueous explosive volcanism, in White, J.D.L., Smellie, J.L., and Clague, D.A., eds., Subaqueous explosive volcanism: American Geophysical Union, Geophysical Monograph, v. 140, p. 1–23.

Wilch, E., and Hughes, T.J., 2000, Calculating basal thermal regimes beneath the Antarctic ice sheet: Journal of Glaciology, v. 46, p. 297–310, doi: 10.3189/172756500781832927.

Wilch, T.I., and McIntosh, W.C., 2002, Lithofacies analysis and $^{40}Ar/^{39}Ar$ geochronology of ice—volcano interactions at Mt. Murphy and the Crary Mountains, Marie Byrd Land, Antarctica, in Smellie, J.L., and Chapman, M.G., eds., Volcano—ice interaction on Earth and Mars: Geological Society of London, Special Publication, v. 202, p. 237–253.

Wilding, M.C., Smellie, J.L., Morgan, S., Lesher, C.E., and Wilson, L., 2004, Cooling process recorded in subglacially erupted rhyolite glasses: Rapid quenching, thermal buffering, and the formation of meltwater: Journal of Geophysical Research, v. 109, doi: 10.1029/2003JB002721.

Wilson, L., and Head, J.W., 1994, Review and analysis of volcanic eruption theory and relationship to observed landforms: Reviews of Geophysics, v. 32, p. 221–263, doi: 10.1029/94RG01113.

Wilson, L., and Head, J.W., 2002, Heat transfer and melting in subglacial basaltic volcanic eruptions: Implications for volcanic deposit morphology and meltwater volumes, in Smellie, J.L., and Chapman, M.G., eds., Volcano—ice interaction on Earth and Mars: Geological Society of London, Special Publication, v. 202, p. 5–26.

MANUSCRIPT ACCEPTED BY THE SOCIETY 3 NOVEMBER 2008

Megascale processes: Natural disasters and human behavior

Susan Werner Kieffer*
Department of Geology, University of Illinois, Urbana, Illinois 61801, USA

Paul Barton
U.S. Geological Survey, M.S. 954, Reston, Virginia 20192, USA

Ward Chesworth
Land Resource Science, University of Guelph, Guelph, Ontario, N1G 2W1, Canada

Allison R. Palmer
4875 Sioux Drive, #206, Boulder, Colorado 80303, USA

Paul Reitan
Department of Geology, State University of New York at Buffalo, Buffalo, New York 14260, USA

E-an Zen
Reston, Virginia 20191, USA

ABSTRACT

Megascale geologic processes, such as earthquakes, tsunamis, volcanic eruptions, floods, and meteoritic impacts have occurred intermittently throughout geologic time, and perhaps on several planets. Unlike other catastrophes discussed in this volume, a unique process is unfolding on Earth, one in which humans may be the driving agent of megadisasters. Although local effects on population clusters may have been catastrophic in the past, human societies have never been interconnected globally at the scale that currently exists. We review some megascale processes and their effects in the past, and compare present conditions and possible outcomes. We then propose that human behavior itself is having effects on the planet that are comparable to, or greater than, these natural disasters. Yet, unlike geologic processes, human behavior is potentially under our control. Because the effects of our behavior threaten the stability, or perhaps even existence, of a civilized society, we call for the creation of a body to institute coherent global, credible, scientifically based action that is sensitive to political, economic, religious, and cultural values. The goal would be to institute aggressive monitoring, identify and understand trends, predict their consequences, and suggest and evaluate alternative actions to attempt to rescue ourselves and our ecosystems from catastrophe. We provide a template modeled after several existing national and international bodies.

---

*skieffer@uiuc.edu

INTRODUCTION: FROM EARTH ISLANDS TO PLANET EARTH TO ISLAND EARTH

Unlike the other planets discussed in this volume, Earth has human inhabitants, and in this paper we look at our own capacity to create a megadisaster comparable to exogenous and endogenous events discussed in other papers. We who have lived through the second half of the twentieth century have witnessed a profound transition in the relation between the physical planet and its human inhabitants. In the middle of the century, the planet still had real islands, both physical and sociological, beyond which were frontiers that held new lands, mysteries, adventures, cultures, and resources. Expanding population and technology merged these islands into a relatively seamless planet by the end of the century.

Astronauts in space took hauntingly beautiful photographs at a resolution that revealed a beautiful and pristine planet Earth, the "Pale Blue Dot" of Carl Sagan (1994) (Fig. 1). However, in the early years of the twenty-first century, we see higher resolution views of this planet that reveal global-scale changes caused by our species. On 4 August 2005 Commander Eileen Collins and her international crewmates aboard the shuttle Discovery said, in a video-conference from space with the Prime Minister and high officials from Japan, "Sometimes you can see how there is erosion, and you can see how there is deforestation. It's very widespread in some parts of the world. We would like to see, from the astronauts' point of view, people take good care of the Earth and replace the resources that have been used. The atmosphere almost looks like an eggshell on an egg, it's so very thin. We know that we don't have much air, we need to protect what we have."

Although life forms have affected the planet on a global scale throughout geologic time (e.g., causing the formation of the oxygen-rich atmosphere), never before has a species been able to observe its effect on its own environment at this scale. This planet is the home, the only home, of our species, and there are many indications that we are altering that home in ways that will undermine our survival and evolution into the civilized global society that we might become. "It has often been said that, if the human species fails to make a go of it here on Earth, some other species will take over the running. This is not correct. We have, or soon will have, exhausted the necessary physical prerequisites so far as this planet is concerned. With coal gone, oil gone, high-grade metallic ores gone, no species however competent can make the long climb from primitive conditions to that high-level technology. This is a one-shot affair. If we fail, this planetary system fails so far as intelligence is concerned" (Hoyle, 1964).

As implied in Hoyle's quote, many of the value-laden comments in this paper have time scales associated with them. For example, if we simplify resources into two categories, renewable and nonrenewable, we need to ask "over what time scale?" All resources may be renewable on geologic time scales, but this is not relevant to the time scales of human needs. We are specifically looking at problems that are urgent on the time scale of collapse of civilized societies. This time scale is decades to centuries.

At the end of 2004, the Sumatran earthquake gave us a glimpse of the enormous megascale geologic processes that can happen on the Earth on human time scales—earthquakes and tsunamis in this case, and floods, hurricanes, landslides, avalanches, volcanic eruptions, and meteorite impacts as discussed elsewhere in this volume. Megascale geologic events are rare by human reckoning, and most occurred in the past when the total human population was small or nonexistent. Earlier human societies were relatively isolated, either on real islands or in local societies, on continents that were effectively isolated from one another by distance and primitive technologies. The effect of a geologic event on individuals could always have been catastrophic, but the catastrophe was not global in scale.

At present, however, societies of the whole planet are so interconnected that planet Earth is essentially a single island, perhaps more aptly, a spaceship. The only remaining islands for us are other planets. We do not have the technology to move from this island to another, and may not have it in the near-enough future. With the present daily net population increase on our planet exceeding 250,000 we will surely never have a means of mass migration for a significant fraction of Earth's population (Hardin, 1993, p. 9–11).

By geologic time scales, human population and the scale and rate of human exploitation of nonhuman resources are exploding so that even rare events may critically affect our survival. In other words, our civilization has become so globally connected that even relatively "small" megascale events on the geologic scale can have potential for immense consequences to our species.

Civilization is a fragile enterprise: we depend on a favorable global climate, abundance of natural resources, and geologic as well as social stability. This fragility is compounded

Figure 1. Our "Pale Blue Dot," the view of the rising Earth that greeted the Apollo 8 astronauts as they came from behind the Moon after the lunar orbit insertion burn. Earth is about 5° above the horizon in the photo. Photo from NASA.

by our propensity to take for granted our planetary resources—and each other—with our current political systems and a short-sighted view of the future.

What has been the effect of past natural megadisasters on humans? What might be the consequences of the same events with the current population distribution? What is the magnitude of the human endeavor compared to these natural events? What can be done to protect humans and the ecosystems on which they depend from disasters? This paper explores these questions.

THE EFFECTS OF NATURAL MEGASCALE EVENTS ON HUMANS: PAST, PRESENT, AND FUTURE

As discussed throughout this volume, earthquakes, volcanoes, tsunamis, and exogenous events such as meteorite impacts or solar activity have the potential to produce megascale catastrophes.

Earthquakes

The deadliest earthquakes in recorded history have occurred in China: in 1556, Shansi, with ~830,000 dead; and in 1976, Tangshan, Hebei, with 255,000 officially known dead and as many as 655,000 possibly dead. Earthquakes in Iran, Syria, Japan, Turkmenistan, Italy, Pakistan, Peru, and Portugal have killed 50,000–200,000 people repeatedly throughout recorded history. By comparison, the M 8.25 San Francisco earthquake in 1906 caused 700–3000 deaths out of the population of 400,000. The Sumatran earthquake was the greatest megascale event recorded by modern technology: the magnitude (M) 9.1–9.3, duration (>10 min), and recorded fault break (1200–1300 km) were the greatest in history (Lay et al., 2005). The tsunamis from the Samatran earthquake were not particularly large on a geologic scale, with a maximum height on the order of 10 m, but because of the density of population in vulnerable areas, 283,000 people died suddenly.[1] Images of the devastation of a megaevent were seen around the whole world, and this quake did not occur in the densely populated economic centers of Indonesia where the death toll could have been much higher. The Kashmir/Pakistan earthquake of 8 October 2005 killed nearly 100,000 people and left 3.3 million homeless to face dying in a brutal winter. However, in both cases, the consequences to humans may have been the most severe because of the secondary events, what we have called the "disasters within the disaster."

There is much current interest in clusters and long-distance connections of geologic processes such as earthquakes and volcanism. Earthquake "storms" pose large-scale and prolonged dangers. A storm is an unusual cluster of very strong earthquakes in a contiguous region spanning several decades, the clusters often separated by centuries of relative quiescence. Compelling evidence indicates that earthquake storms caused the demise of the Bronze Age civilizations in the eastern Mediterranean ~1225–1175 B.C. (Nur and Cline, 2000). The civilization failed by a system collapse—the destruction of buildings, massive migrations of people, loss of culture, famine, and uprisings. At this time, between 2000 and 1000 B.C., the world population is estimated to have been in the range of 25–50 million, and the population in the Middle East was possibly between 6 and 9 million.

In comparison, the combined population of the Middle East and North Africa was 300 million people in 2001, and is projected by the World Bank to approach 388 million by 2015. One population center at particular risk in modern times is Istanbul. More than 10 million people live in Istanbul alone today, more than in the whole Middle East 3000–4000 yr ago. The largest earthquakes in Turkey have occurred on the North Anatolian fault. An earthquake storm appears to have started in 1939 with a M 7.8 earthquake near Erzincan. Earthquake activity migrated westward through 1969, rupturing the fault zone in a series of earthquakes with M 7 and greater (Okumura et al., 1993). The fault geometry becomes complicated at the western end, but two M 7+ earthquakes occurred there in 1999, killing over 18,000 people and causing $25 billion in damage. U.S. Geological Survey scientists (Parsons et al., 2000) have estimated that there is a 60% chance that a large earthquake will hit Istanbul by 2030. Many other major cities around the world are subject to similar earthquake risks. Earthquakes (and potentially bolide impacts, see below) near or in bodies of water pose an even greater danger because of the potential for tsunamis.

Civilization in the Mediterranean areas has been affected more than once by megascale events such as earthquakes, volcanic eruptions, and tsunamis. Strong earthquakes in the seventeenth century B.C. destroyed many Minoan palaces on Crete. These were followed by the eruption of ~30 cubic km of magma from Santorini in ca. 1627–1628 B.C. (Grudd et al., 2000). The Minoan city of Akroteri was smothered under ~2 m of ash, and the ash deposits destroyed the agricultural fields. These events are believed to have contributed to, or caused, the end of the Minoan civilization centered on the island of Crete, again probably by system collapse. It has been suggested that tsunamis from the eruption 40 km away not only destroyed cities, but also destroyed the navy at Crete. Over the next few centuries, the navy lost crucial battles with the Mycenaean navy, so that a former colony took over the empire.[2]

Hurricanes

Hurricanes, also called typhoons or tropical cyclones, are events that have occurred throughout recorded history, but only in the past few decades has there been sophisticated instrumentation to allow us to describe them accurately. Because of this relatively short time of monitoring, the "biggest" events recorded to date are likely not to have been the biggest that have occurred.

The intensity of a hurricane is measured by either the low pressure in the center or the maximum sustained winds at ground

---

[1]Earthquake statistics used in this paper are from http://neic.usgs.gov/neis/eqlists/eqsmosde.html.

[2]An accessible encyclopedia article on this is http://www.nationmaster.com/encyclopedia/Bronze-age.

level. Hurricane Katrina's lowest pressure was 902 millibars (mb), and maximum sustained winds were ~150 miles per hour (mph). (By comparison, hurricanes in different meteorological settings can have different properties: the most intense hurricane recorded was Typhoon Tip in the Northwest Pacific Ocean, 12 October 1979, with a much lower central pressure of 870 mb, and maximum sustained winds of 190 mph.)

Katrina intensified over a period of four days after it entered the Gulf of Mexico in late August 2005, providing time for planning and evacuation. Lessons learned from Katrina were implemented when Hurricane Rita came into a nearby region in late September 2005. Other tropical storms in other oceanographic settings have been even more severe: Typhoon Forrest in September 1983 intensified in just under 24 h as the pressure dropped from 1000 mb to 876 mb. Estimated surface sustained winds increased 35 mph in just 6 h, and 99 mph in one day, reaching 190 mph. Katrina produced a storm surge on the order of 20 ft; the Bathurst Bay hurricane in Australia in 1899 produced a 42-ft storm surge. Tropical Cyclone Denise dropped 97 in. of rain on La Reunion Island in 1958; Cyclone Hyacinth dropped 128 in. of rain on the same island in just 3 days in 1980, and 223 in. in the 10 days of the storm duration.

Tsunamis

Earthquakes, and potentially bolide impacts (see below), near bodies of water often cause tsunamis. In ~900 A.D., an earthquake on the Seattle fault in the present state of Washington, USA, sent a tsunami throughout the waters of Puget Sound, burying Native American fire pits beneath sand that was swept ashore by the wave (Koshimura et al., 2002; Atwater and Moore, 1992). A recent interpretation of historical Japanese documents, and computer simulations, suggest that an ~M 9 earthquake occurred on the same fault between 1680 and 1720 A.D. (Atwater, 1992; Yamaguchi et al., 1997). The populations of Puget Sound in 900 or 1700 A.D. are unknown, but would have been nowhere near the 3 million people that live and work in this area today. If an earthquake of comparable magnitude should occur in the colliding tectonic plates of the northwestern United States or southern British Columbia, Canada, the shaking, especially of tall buildings, could extend from the heavily populated areas of Vancouver, British Columbia to northern California and could last for several minutes causing extensive destruction. An earthquake of M 9 could also send tidal waves as large as 11 m onto shore in the Northwestern United States within minutes, and send additional waves westward across the Pacific Ocean.[3] Such quakes occur on average once every 500 yr: the last big quake having occurred ~300 yr ago (the fault is currently locked, accumulating energy for a future destructive event). Intensive monitoring and warning efforts may minimize damage and loss of life, but there will inevitably be economic repercussions around the globe from the disruption of such a densely populated, high-technology area.

Volcanoes

Seattle is not only in danger from tsunamis, but also from a future eruption of Mount Rainier, which is volcanically active. It has a magnificent and beautiful summit cap of ice. However, intrusion of magma into the edifice of the volcano could cause melting of the ice, resulting in the generation of enormous mudflows like those at Nevado del Ruiz, Colombia that killed 29,000 people within minutes in 1985. A mass of mud, ice, and water traveled a distance of 100 km down the White River valley 5600 yr ago into what is now the middle of Seattle and Tacoma (Valance and Scott, 1997). This event covered the land with a layer of mud 90 m thick in places. Large mudflows have traveled the same path on the average of 600 yr, and over 100,000 people live directly on the young mudflows.

Volcanoes also emit gases (primarily SO_2, H_2O, and CO_2) that alter the composition of the atmosphere. Approximately 73,000 yr ago, the volcano Toba in Indonesia erupted ~2800 km^3 of magma, a volume that only seven seas or lakes in the world surpass (Caspian, Baikal, Tanganyka, Superior, Nyasa [Malawi], Michigan, and Huron). The ash and gases from the eruption caused short-term global cooling of 3–5 °C, and perhaps as much as 15–20 °C in local regions (Rampino and Self, 1992, 1993). The eruption coincided with, and perhaps caused, a reduction of human population to fewer than 3000–10,000 individuals at that time (Ambrose, 1998). Historically, even smaller eruptions have produced temporary coolings that have had mild to severe impacts regionally. Examples include eruptions from: Laki (1883, 4.8 °C cooling in Europe, ~1 °C in the eastern United States); Tambora (1815, snow in all summer months in New England and Europe); and Pinatubo (1999, world temperature decrease of ~1 °C over 2 yr). The eruption of Laki was tiny compared to eruptions that we know have happened in the past and that have had similar chemistry—flood basalts. The significant impact of its sulfurous emissions on climate suggests possible catastrophic effects of eruptions on ecosystems and humans.

Exogenous Events

We have thus far considered events endogenous to the Earth, but we should be aware that there are also important exogenous events, viz., large solar outbursts of charged particles, supernova explosions, and meteorite impacts. The effects of an impact of a meteorite ~10 km in diam 65 Ma ago caused the extinction of a large number of animal and plant species, including the dinosaurs (Alvarez et al., 1980). We do not need meteorite impact events of this scale to cause major global disruptions at the present levels of population. Added to the direct-hit effects—including atmospheric shock waves and tsunamis—there would be secondary effects on the climate and ecological support systems, and tertiary effects on dependent humans and a functioning global civi-

---

[3]An updated summary of Northwest tsunami hazards is available at http://www.pmel.noaa.gov/tsunami/time/or/resources.shtml.

lization, another scenario for system collapse. Solar flares have caused significant economic disruption within the past decade by disrupting modern communications. Our economic systems and our survival depend on the resilience of our response to these uncontrollable processes.

The megascale processes described above generally cannot be prevented. Thus far, we cannot tell which of the numerous possible events will occur first, or when. We simply know from our geologic concept of "deep time" (Palmer, 2000) that these events will occur. In some instances, mitigation and remediation are possible—for example, human casualties of the tsunami in the Indian Ocean could have been mitigated by an effective warning system. A deep space warning system may give us advance notice of a meteorite heading toward us, just as volcanic monitoring systems may give us warning of new activity. However, mitigation costs money, and mitigation of rare events is traditionally given a lower priority than mitigation of frequent events. We live on a planet with constantly changing geological conditions: some are gradual, some are episodic, and both can be of a scale to be catastrophic.

HUMAN BEHAVIOR AS A MEGASCALE PROCESS

There is an on-going megascale process not yet considered. It is unique in the history of the Earth: the expansion of an animal species with a population and brain large enough to challenge all competition in the ecosystem—humans. Other so-called "terminator species" (Flannery, 1994) have existed in the past, but never at the global scale. We have been able to use our brains to bypass, or delay, the negative feedback of natural selection—the process that keeps other species from dominating the biosphere. Consequently, the size of the human population has increased along an exponential trend. For perspective, the deaths in World Wars I and II, and those from the Spanish flu epidemic during World War I, did not produce a noticeable dent in the population growth curve. At current birth rates, over 250,000 babies were born in slightly less than 1 day after the 26 December 2004 earthquake that took a similar number of lives. We are the only big fierce animals (Colinvaux, 1979) remaining whose population is growing.

Population

Exponential population growth means that our behavior en masse, particularly consumption of resources and generation of waste, has also been magnified along an exponential trend. Stripped to its fundamentals, human behavior is no different from the behavior of other animals: we eat, reproduce, and ignore our dependence on other components of the ecosystem at our own peril just as do rabbits, cockroaches, foxes, and lemmings.

The number of humans that the planet could support, the so-called "carrying capacity," is debatable. Extremes vary from 0.5 to 14 billion; medians of the low and high estimates yield a range from 2.1 to 5.0 billion (Cohen, 1995). This number depends strongly on assumptions about standards of living, technology, resources, and whether or not we live in a world of strongly recycled or vastly depleted natural resources. Our current population exceeds 6 billion and is projected to reach 9 billion by 2050. For this paper, we arbitrarily consider a future in which 3 billion people might live sustainably.

The major difference between "us" and "the rest" of the ecosystem is that we have been able to extend our lifetimes and avoid some of the obvious natural selection processes. The fact that we have probably overshot the carrying capacity of the planet with respect to our species means that this "natural" behavior has become problematic to our own survival. Collective human behavior is affecting the physical and biological state of the planet on a massive, and dramatically rapid, scale. Many tend to view the effects of human behavior on the planet as gradual because of the perception of time on a human, rather than geologic, time scale. However, all island societies have collapsed within geologically short time scales when the populations exceeded the carrying capacity. We are heading toward a megascale catastrophe on island Earth if we cannot change our behavior (Diamond, 2004; Wright, 2004).

Suppose, for example, that when the Earth population reaches 9 billion people in ~2050, a natural event of the type described above caused catastrophic system collapse. Suppose that conditions and resources after the collapse permitted only 3 billion people to ultimately survive. Suppose that this collapse lasts 60 yr. The result would be 250,000 more deaths per day than births per day for 60 yr, i.e., the magnitude of the deaths from the 26 December 2004 tsunami would be repeated daily for 60 yr!

However, this example treats the human conditions around the world as uniform. They are not. Both within and between countries, the rich are getting richer and the poor are getting poorer. The United States has less than 5% of the world's population, consumes one-quarter of its resources, and generates at least that fraction of the world's waste. It has been estimated that each child born in the United States will consume 10–15 times the resources of a child born in India (Wackernagel and Rees, 1996, p. 85, Table 3.4 using 1991 data; Zen, 2000, p. 389–390).

Thus the standard U.S. family with two children is equivalent to a family of 20–30 children in the Third World. Each American consumes more grain per year than 150 Bangladeshis or 500 Ethiopians. This disparity, on the other hand, exposes the high-technology, consuming nations to greater vulnerability to a megadisaster than, for example, subsistence farmers because of the dependence on a functional global economy.

Another aspect of the current population distribution is the vulnerability of food production and distribution networks. Consider the grain situation in the world. Much of the world's grain comes from prairies of the mid-continents of the United States and Canada. Upwind from the plains is the active volcano of Yellowstone, Wyoming, currently being monitored for potential large-scale volcanic activity. An eruption occurred at Yellowstone 2 Ma ago comparable to the Toba eruption mentioned before. Only 600,000 yr ago, the volcano erupted more than 1000 km^3 of ash. If Yellowstone had a volcanic event of these magnitudes today, roughly half of the grain-producing area of North America would be covered with more than 10 cm of volcanic ash (Fisher et al., 1998), and a system collapse would inevitably follow.

The development of human society has already led to a range of well-recognized man-made hazards, which, if they occurred, could be comparable in scale to some of the natural megaevents. Some are obvious: for example, nuclear warfare, bio- or technological terrorism, nanotechnology gone wild, a volcanic intrusion into a nuclear waste repository, or the strange scenario of a high-energy physics experiment gone awry (see Posner [2004] for a discussion of risk and response to rare, but high-consequence events such as these). As serious as these events may be, our integrated collective behavior on a much less obvious scale poses equal or perhaps even more dangerous threats. All but the most extreme of these man-made hazards will not impact the global environment on the scale of our collective behavior. There is clear evidence from observation of many components of the planet that our activities are changing the environment on which we depend, in ways that will preclude traditional future use: soils, rivers, climate, water, landscape, resource distribution, and ecosystems. A few examples illustrate the wide range of our impact.

Rivers

For over 7000 yr, the manipulation of the Nile River by humans has affected increasing areas of the river corridor, its delta, and now, the Mediterranean Sea (Stanley and Warne, 1998). During much of this history, the population of all of Egypt (concentrated mostly in Memphis on the northern part of the Nile, and Thebes in the southern highlands) was fewer than 1 million people; but even as early as 2200 B.C., limits of the resources on the Nile seem to have caused the decline of the Old Kingdom.

The alteration of the Nile has continued, accelerating dramatically in the nineteenth and twentieth centuries with the building of the Aswan Dams from 1889 to 1970. Today, ~60 million people live on the delta, with population densities of 1000 people/km^2, and much greater in places like Cairo. Nearly all 2500 km^2 of the delta have been affected by the manipulations of the Nile (Farvar and Milton, 1972). The river no longer cleanses its banks or deposits new soil. This change in the river has amplified the effects of humans, e.g., artificial fertilizer, now necessary because the soil is not replaced, leaves deposits of toxic minerals. Human-induced conditions of stagnant water have caused a resurgence of severe diseases. Loss of silt and algae from the Nile has caused the disappearance of a major sardine resource from the Mediterranean and its shrimp population has been adversely affected by the reduced discharge of the Nile. Loss of the fresh water cover is causing invasion of Red Sea fauna into the Mediterranean across the Suez Canal.

Hydrologic Cycle and Climate

On a larger scale, humans have transformed the global hydrologic cycle, partly caused by human-abetted global warming, and evidenced by retreat of glaciers and thinning of Arctic ice (see IPCC [2001,Table 3–2, p. 70] for predictions of events absent climate policy interventions). There is high confidence that by 2025 there will be further retreat of glaciers, decreased sea ice, thawing of some permafrost, and longer ice-free seasons on rivers and lakes, and medium confidence that there will be extensive sea-ice reduction (Kennedy and Hanson, 2006). Global warming will increase the severity of hurricanes, an indication perhaps being hurricanes Katrina and Rita in 2005.

We have changed the climate and even the solar input to the surface of the Earth through the formation of the ozone hole. The evidence that the climate is changing, and that human contributions now are overriding the natural variability, has become compelling to most scientists (Karl and Trenberth, 2003).

Carbon is cycled through the biosphere by both biological processes (dominant) and geologic processes, in the form of CO_2 and other gases, in solution in water, and bound in carbonate rocks. Time scales vary considerably, with volcanism and silicate weathering playing significant roles on geologic time scales, but small roles on human time scales. Photosynthesis accounts for 120 Gt/yr of carbon (metric gigatonnes, GtC), about half through respiration and half through net primary production.[4] On a global scale, humans, who represent roughly 0.5% of the total heterotroph biomass on Earth, appropriate ~32% of the net primary production (Imhoff et al., 2004). Total anthropogenic emissions from fossil fuel burning, cement production, and changes in tropical land use are ~7 GtC/yr (Schimel et al., p. 76–86 in IPCC, 2001). How much is 7 billion tonnes of carbon? An average human weighs ~0.075 tonne. So, this mass of carbon is equal in weight (not in carbon mass) to ~93 billion people, or 15 times the entire population of the planet. About 45% of the anthropogenic CO_2 stays aloft in the atmosphere. Even if we allow for controversy about the human contribution to this change by our behavior, these numbers suggest that we should act prudently.

Sulfur from natural sources (volcanoes, biogenic marine, and terrestrial biogenic) is still a significant fraction of the total sulfur balance in the tropical latitudes of the northern hemisphere and in all latitude belts of the southern hemisphere. However, between 35° and 50° N lat, only 8% of the sulfur emissions come from natural sources, and the human-produced emissions dominate the global sulfur budget (see Crutzen et al. [2003] for a general discussion of the effect of "parasols" on climate).

Other large human impacts include our creation of megacities where the paving of so much land surface has altered the albedo, the hydrology, and the pattern of heat from solar input. We have changed the definition of "dark night side of the Earth" by the artificial illumination of our cities. Within the continental United States, Mexico, Japan, and China, surface temperatures on weekends (Saturday–Monday) vary systematically from weekday weather (Wednesday–Friday) over large spatial scales

---

[4]Net primary production is the net amount of solar energy converted to plant organic matter through photosynthesis, usually measured in units of elemental carbon. This is the primary food energy source for the global ecosystems.

(Forster and Solomon, 2003). The effect can be as large as 0.5 °C, but whether this is an increase or a decrease of temperature is not the same in all locations. Aerosol-cloud interactions caused by accumulated man-made products are suspected as the cause.

Soils, Energy, and Minerals

We have altered soils through compaction and erosion with unsustainable agriculture practices. Human-caused soil erosion rates exceed those of steady geologic processes: we move earth at a rate of ~35 Gt/yr, ~3 times that of all other natural agents, mostly through plowing (Hooke, 1994). Plowing, overgrazing, compaction, acidification from acid rain, and the use of fertilizers and biocides are changing the physical, chemical, and biological nature of soil. Numerous studies show that worldwide, soil erosion is occurring much faster than renewal rates (Pimentel et al., 1995; Pimm et al., 1995). Soil cannot be viewed as a renewable resource on the time scale of humans. That it has been so regarded may account for the fact that our record in soil conservation is so abysmal. Soil is necessary not just for agriculture but also for the entire terrestrial ecosystem: no soil, no terrestrial plant life, no land-based photosynthesis, and little food for animals or humans. Unless we learn to use the soil sustainably, we will never be able to maintain the food supply that our complex society depends upon (Leopold, 1949).

We have extracted energy and minerals at such a rate that scarcities are likely to occur within decades, at worst, or centuries, at best. Current estimates are that peak production of liquid petroleum and natural gas will occur before 2050, possibly decades earlier. Even a new rare, 10 billion barrel supergiant field, if all could be extracted, would delay peaking by only a few months or very few years (Campbell, 1997; Duncan and Youngquist, 1999; Gluskoter, 1999). With population increasing toward 9 billion and increasing per capita energy demand, we could exhaust coal supplies within centuries. Because of high CO_2 levels and soot pollution from coal—the question is—"Should we want to burn this coal, even if we can?"

Referring back to our Introduction in which we emphasized that our concerns are about time scales that affect the stability of civilized societies, we argue that neither fossil fuel energy sources nor soils, or possibly some critical mineral resources, are renewable on relevant time scales. It could be argued that our waste dumps or even the wasted remains of our cities are essentially economic resources for a future civilization because we have done a huge amount of work concentrating minerals as we have used up our fossil fuels; or, perhaps some soils could be considered "enriched" for future ecologies rather than viewed as "poisoned" for our own ecologies. In our view, such arguments lie in the realm of science fiction or fantasy rather than science fact or rational philosophy.

The Rest of the Ecosystem

As the current terminator species, we are causing the extinction of other species on a massive scale. Estimates of extinction rates are extremely variable and difficult to make because the time scales for ecology are so short compared to those of geology. Background rates of extinction over 600 Ma are estimated to have been ~1–10 species/yr for every million species on the earth, or 0.0001% per yr. Estimated current rates of extinction by a variety of methods are ~1000 times prehuman levels (0.1% per yr) with the rate projected to rise, possibly very sharply (Pimm et al., 1995). Although the concept is qualitative, comparisons have been made of the present situation to the five times in Earth history that have been "mass extinctions" in which more than 50% of species went extinct in a geologically short period of time.

The impact of the extinctions is already being felt in the economies of the world, and in food shortages affecting human welfare. We have removed the top predators from a number of important systems resulting in cascading effects through ecosystems. Frank et al. (2005) have documented this effect in the decline of the large cod-dominated marine ecosystem of eastern Canada. The trophic system becomes completely restructured, at least on time scales significant to economics. It is possible that this is not just a situation of dwindling numbers that could recover, but of extinctions.

The examples given so far relate to the active role of humans in bringing about disaster by our collective behavior in consumption and waste production. The sheer size of the human population means that we may also be the passive cause of disaster. Returning to the Sumatran tsunami, one could cogently argue that the loss of 150,000–300,000 lives in southern Asia was caused by overpopulation that forced people to live in vulnerable areas, in combination with a blatant disregard for the preservation of a sustainable ecosystem.

There is one major, and potentially hopeful, difference between the random catastrophic event and the gradual build up to a potential human-induced catastrophe: namely, we do have advance warning of the approach of a disaster, as shown by the examples cited above. We ignore this warning at our peril. Encouragingly, the megascale process of the evolution of our species is unique in being within the grasp of human control—if we can muster the collective will to act on the evidence. Given that we humans have the ability to think, act, and organize, we have the potential to moderate our own behavior. Unfortunately, we are reluctant to fully acknowledge our abuse of our air, water, and soils, the consequences of our unfettered energy and mineral consumption, and our contempt for the rest of the ecosystem. Due to our own survival instincts, reinforced by both religious and political thought that we are the masters of the domain rather than merely part of it and an economic belief that we cannot sacrifice short-term gain, we maintain a state of denial about our effect on the planet. We seem curiously unwilling to take the actions needed to rescue ourselves from that self-induced megascale terminal event called "extinction of civilized society."

FUTURE OPTIONS IN A COMPLEX SYSTEM

Civilization is a complex system, in the strict sense in which that concept is emerging in modern mathematics (Bar-Yam,

2002). An inherent property of complex systems is that there is some parameter that is "tuned" until the unexpected happens. For a thought experiment, consider observing a pot of water over a fire and pretend that you had never seen water heated in this way before. The temperature of the water is being "tuned" by the heat from the fire. Nothing much happens as the temperature rises from ambient, except maybe the formation of a few bubbles as dissolved air comes out of solution. However, as the temperature approaches 100 °C, the "unexpected happens": the water begins to boil, it suddenly changes state from a single-phase liquid to a two-phase boiling mixture and, ultimately, to vapor.

A major tuning parameter for populations is the number of members—if the number is increased relative to critical resources, nothing dramatic happens for some time. Then the resource is rather suddenly depleted and the population declines or collapses. The system typically reverts back to a less complex structure, and indeed, the loss of complexity in structure has been used as a definition of the collapse of societies (Tainter, 1988).

A complex system can be organized in various ways. It can be chaotic (resembling a random system), or it can be in different stable states. To qualitatively follow the boiling water analogy above, water can be in the stable states that we call ice, liquid, or vapor; or it can be chaotic, i.e., on the phase boundaries between solid and liquid, or liquid and vapor. A challenge to our civilization is to find a stable state. Because we have depleted nonrenewable resources, we do not have the choice of going back to a previous stable state, but must search for a new relationship between the human population and the planet on which we live.

Organization can change the nature and operation of the system. Evidence from the past is that human civilizations have not been able to organize coherently to prevent overuse of resources (Tainter, 1988, 1995; Diamond, 2005). The current global impasse over acknowledging the magnitude of the problems and negotiating ways to intervene (viz., the Kyoto Treaty process) is an example of continuing lack of coherent organization. In complex systems, partial organization does not put the system into a stable state; the whole system must be "self-organized." The implication is that in this world of global interdependence, the next stable state must include the whole planet—both its peoples and its resources.

This same concept was stated more broadly by the environmental philosopher J. Baird Callicott (2005), in reviewing the seminal paper on environmental ethics by Lynn White, Jr. (1967). White's fundamental assumption was that "what we do collectively depends on what we collectively think," which leads to the conclusion that "if we are to change what we do to the environment, we must begin by changing what we think about the environment." Education and public awareness of the true state of the world is imperative, to be followed by action on that knowledge (Reitan, 2005).

Two options or courses of action appear to be open to us: (1) Do nothing or little, and wait for the catastrophe to occur, which is the current state and is deemed most likely by Milbrath (1989, p. 340). (2) Attempt remediation and conservation, as fast as is humanly possible, which we will call "adaptation." In either scenario, we need to prepare for adaptation to an inevitably rapidly changing world. Hence we come to the final point of this paper: We need to understand that we humans are a geologic force affecting the whole planet, and we need to break the collective denial about our effects on the planet and institute aggressive procedures to monitor, understand, evaluate, predict, and recommend actions for the health of the planet.

During the time we have been thinking about this problem the epidemic of sudden acute respiratory syndrome (SARS) and the potential epidemic of bird flu have dramatically illustrated the global interconnectedness and fragility of our species. Without rapid detection, acknowledgment of the problem, aggressive monitoring, and treatment through such organizations as the Centers for Disease Control (CDC) and the World Health Organization (WHO), catastrophic pandemics of SARS or flu might have occurred (and still may). Monitoring showed that in the absence of initial control measures, individual SARS spreaders infected 2.7 or 3 people on average in Hong Kong and Singapore, respectively (Riley et al., 2003; Lipsitch et al., 2003). As control measures were instituted, the transmission rate fell. With four key actions—detection, acknowledgment, aggressive monitoring, and treatment—the pandemic was at least temporarily halted in the societies that instituted these remedial measures.

By analogy, thoughtful people have already detected and defined many aspects of human behavior that are jeopardizing the planet. Regardless of which of the two paths we follow (do nothing, or remediate and adapt), a large body of wise leadership, and a huge database for research, education, and policy will be required at a global scale (Zen et al. [2002] proposed a global monitoring system as part of a prudent strategy for using Earth's resources wisely). For example, the very concept of "adaptation" of humans to the natural world in the midst of an ecological catastrophe is poorly defined, if defined at all. The resources for analyzing and thinking of the world on a global scale do not currently exist (Speth [2004], and in previous works, discusses this in detail; see also Ehrlich and Kennedy [2005]). We examined the charter of the United Nations and found aspects of its mandate to be appropriate, but because of its historical and current focus explicitly on war and war prevention, the focus is only obliquely on environmental sustainability (e.g., it is only one of eight of the Millennium Development Goals).

We realize that a number of different organizations will, in fact, be required to deal with such a massive problem, e.g., a scientific body to provide impartial facts and uncertainties, an engineering body to propose and implement technical solutions, a negotiating body to balance the realities of political, economic, religious, and cultural values (like the United Nations, and like the new initiative on human behavior advocated by Ehrlich and Kennedy [2005]), and an enforcement body that is responsive to all of the inputs (like the Canadian peace-keeping forces?). However, some entity must be seen as the best provider of scientific truth and reality, as free from bias as is humanly possible. Current efforts of global collaboration by the National Academies[5] represent a positive development. However, given the volunteer status of most members to the Academies, and the lack of policy that mandates that they be at the political and moral bargaining tables, much needs to be done.

Freedom from bias in the duties of a scientist does not imply lack of responsibility to provide the best estimates of consequences of actions and the uncertainties, nor opinions about what will and will not work for humankind. This kind of scientific assessment can be done independently of political, economic, religious, and cultural pressures, as long as the judgments about what to do about it include those who deal with human morality, economics, and politics.

To address our perceived global scientific needs, we have taken the mandates of the Centers for Disease Control and modified them to arrive at a concept of a mandate for a much-needed global body, a "CDC for Planet Earth" (CDCPE, perhaps standing for the "Center for Disaster Control for Planet Earth" in our thinking).

PROPOSAL FOR A CDCPE

The existing CDC mandate is

The Centers for Disease Control and Prevention (CDC) is recognized as the lead federal agency for protecting the health and safety of people - at home and abroad, providing credible information to enhance health decisions, and promoting health through strong partnerships. CDC serves as the national focus for developing and applying disease prevention and control, environmental health, and health promotion and education activities designed to improve the health of the people of the United States.

To paraphrase the CDC mandate and take it to a global scale, we propose that:

"The CDCPE should be recognized as the lead world body for protecting the long term health and safety of the planet and all of its inhabitants, providing credible information to enhance decisions relating to all resources of the planet, and promoting wisdom in resource use through strong international cooperation. The CDCPE serves as the international focus for developing and applying resource conservation, and promoting education activities designed to improve the conditions for continued human existence on the planet."

In carrying out its activities, the CDCPE might have the following five core functions for resource evaluation:

- *Manage information by assessing resource data and their uncertainties, and assessing trends; set the agenda for, and stimulate, research and development.*
- *Set, validate, monitor, and pursue the proper implementation of norms and standards.*

- *Catalyze change through technical and policy support that stimulates cooperation and action and helps to build sustainable global capacity.*
- *Negotiate and sustain national and global partnerships.*
- *Articulate consistent, ethical, and evidence-based policy and advocacy positions.*

This body needs at the least to be (1) global, (2) credible, (3) scientifically based, and (4) sensitive to political, economic, religious, and cultural values while avoiding direct bias. If we have the collective will to build a "CDC for Planet Earth" to institute aggressive monitoring, to identify and understand trends, to predict their consequences, and to suggest and evaluate alternative actions, we may be able to rescue ourselves and our ecosystems from catastrophe. Such actions would be prudent insurance and necessary if remediation were attempted.

CONCLUSION

We must do something, and we must be sure that the actions are real, and not delusions of action. We agree with the quote from Hoyle at the beginning of this paper, with one exception: perhaps civilization is not a one-shot kick at the can. Something majestic might arise from our rubble using renewable resources alone, and certainly the individuals that would create such a future civilization would have to be collectively wise in ways that we have not been. If we can envision such a future society, can we not envision a way to become that civilization ourselves? Planet Earth—that small blue dot—is now island Earth and it is time for action to ensure our survival on that island.

ACKNOWLEDGMENTS

We appreciate the thoughtful comments of Clark Bullard, Laszlo Keszthelyi, and Mary Chapman in review.

REFERENCES CITED

Atwater, B.F., 1992, Geologic evidence for earthquakes during the past 2000 years along the Copalis River, Southern Coastal Washington: Journal of Geophysical Research, v. 97, no. B2, p. 1901–1909, doi: 10.1029/91JB02346.

Atwater, B.F., and Moore, A.L., 1992, A tsunami about 1000 years ago in Puget Sound, Washington: Science, v. 258, no. 5088, p. 1614–1617, doi: 10.1126/science.258.5088.1614.

Alvarez, L.W., Alvarez, W., Asaro, F., and Michel, H.V., 1980, Extraterrestrial cause for the Cretaceous Tertiary Extinction: Science, v. 208, p. 1095, doi: 10.1126/science.208.4448.1095.

Ambrose, S., 1998, Late Pleistocene human population bottlenecks, volcanic winter, and differentiation of modern humans: Journal of Human Evolution, v. 35, p. 115–118.

Bar-Yam, Y., 2002, Complexity rising: From human beings to human civilization, a complexity profile: Encyclopedia of Life Support Systems (EOLSS) developed under the Auspices of the UNESCO, EOLSS Publishers, Oxford, UK, http://www.eolss.net.

Callicot, J.B., 2005, http://environment.harvard.edu/religion/disciplines/ethics/index.html.

Campbell, C.J., 1997, The Coming Oil Crisis: Essex, England, Multi-science Publ. Co., 210 p.

Cohen, J., 1995, How Many People Can the Earth Support?: New York, W.W. Norton, 532 pp.

---

[5]The United States, National Academy of Sciences, the National Academy of Engineering, the Institute of Medicine, and their global counterparts in other countries.

Colinvaux, P., 1979, Why Big Fierce Animals are Rare: Princeton, Princeton University Press, 264 pp.

Crutzen, P.J., Ramanathan, V., Anderson, T.L., Charlson, R.J., Schwartz, S.E., Knutti, R., Boucher, O., Rodhe, H., and Heintzenbert, J., 2003, The parasol effect on climate: Science, v. 302, p. 1679–1681, doi: 10.1126/science.302.5651.1679.

Diamond, J., 2005, Collapse: How Societies Choose to Fail or Succeed: New York, Viking Press, 591 pp.

Duncan, R.C., and Youngquist, W., 1999, Enhancing the peak of world oil production: Natural Resources Research, v. 8, p. 219–232, doi: 10.1023/A:1021646131122.

Ehrlich, P.R., and Kennedy, D., 2005, Sustainability: Millennium assessment of human behavior: Science, v. 209, no. 5734, p. 562–536, doi: 10.1126/science.1113028.

Farvar, M.T., and Milton, J.P., eds., 1972, The Careless Technology, Ecology and International Development: Garden City, Natural History Press, 1030 p. (available online by chapter).

Fisher, R.V., Heiken, G., and Hulen, J.B., 1998, Volcanoes: Crucibles of Change: Princeton, Princeton University, 334 p.

Flannery, T., 2002, The Future eaters: An ecological history of the Australasian lands and people: New York, Grove Press, 432 p.

Forster, P., and Solomon, S., 2003, Observations of a "weekend effect" in diurnal temperature range: Proceedings of the National Academy of Sciences of the United States of America, v. 100, p. 11,225–11,230, doi: 10.1073/pnas.2034034100.

Frank, K., Petrie, B., Choi, J., and Leggett, W., 2005, Trophic cascades in a formerly cod-dominated ecosystem: Science, v. 308, p. 1621–1623, doi: 10.1126/science.1113075.

Gluskoter, H., 1999, Increase in fossil fuel utilization in the twenty-first century: Environmental impact and lower carbon alternatives: Division of Fuel Chemistry of the American Chemical Society, Preprints, v. 44, no. 1, p. 36.

Grudd, H., Briffa, K.R., Gunnarson, B.E., and Linderholm, H.W., 2000, Swedish tree rings provide new evidence in support of a major, widespread environmental disruption in 1628 BC: Geophysical Research Letters, v. 27, p. 2957–2960, doi: 10.1029/1999GL010852.

Hardin, G., 1993, Living within Limits: Ecology, Economics, and Population Taboos: New York, Oxford University Press, 339 p.

Hooke, R., 1994, On the efficacy of humans as geomorphic agents: GSA Today, v. 4, no. 9, p. 217, and 224–225.

Hooke, R., 1994, Spatial distribution of human geomorphic activity in the United States: Comparison with rivers: Earth Surface Processes and Landforms, v. 24, p. 687–692, doi: 10.1002/(SICI)1096-9837(199908)24:8<687::AID-ESP991>3.0.CO;2-#.

Hoyle, F., 1964, Of Men and Galaxies: Seattle, University of Washington Press, 73 p.

Imhoff, M.L., Bounoua, L., Ricketts, T., Loucks, C., Harriss, R., and Lawrence, W., 2004, Global patterns in human consumption of net primary production: Nature, v. 429, p. 870–873, doi: 10.1038/nature02619.

IPCC, 2001, Climate Change 2001: Synthesis Report. A Contribution of Working Groups I, II, and III to the third Assessment Report of the Intergovernmental Panel on Climate Change, Watson, R.T., and the Core Writing Team, eds.: Cambridge, UK, and New York, Cambridge University Press, 398 p.

Karl, T.R., and Trenberth, K.E., 2003, Modern global climate change: Science, v. 302, p. 1719–1723, doi: 10.1126/science.1090228.

Kennedy, D., and Hanson, B., 2006, Ice and history: Science, v. 311, p. 1673, doi: 10.1126/science.1127485.

Koshimura, S., Mofjeld, H.O., Gonzalez, F.I., and Moore, A.L., 2002, Modeling the 1010 bp tsunami in Puget Sound, Washington: Geophysical Research Letters, v. 29, no. 20, p. 1948, doi: 10.1029/2002GL015170.

Lay, T., Kanamori, H., Ammon, C., Nettles, M., Ward, S., Aster, R., Beck, S., Bilek, S., Brudzinski, M., Butler, R., DeShon, H., Ekstrom, G., Satake, K., and Sipkin, S., 2005, The great Sumatra-Andaman Earthquake of 26 December, 2004: Science, v. 308, p. 1127–1133, doi: 10.1126/science.1112250.

Leopold, A., 1949, Sand County Almanac, and Sketches Here and There: New York, Oxford, 226 p.

Lipsitch, M., Cohen, T., Cooper, B., Robins, J., Ma, S., James, L., Gopalakrishna, G., Chew, S., Tan, C., Samore, M., Fisman, D., and Murray, M., 2003, Transmission dynamics and control of severe acute respiratory syndrome: Science, v. 300, p. 1966–1970, doi: 10.1126/science.1086616.

Milbrath, L.W., 1989, Envisioning a Sustainable Society: Learning Our Way Out: Albany, NY, State University of New York Press, 403 p.

Nur, A., and Cline, E.H., 2000, Poseidon's horses: Plate tectonics and earthquakes storms in the Late Bronze Age Aegean and Eastern Mediterranean: Journal of Archaeological Science, v. 27, p. 43–63, doi: 10.1006/jasc.1999.0431.

Okumura, K., Yoshioka, T., and Kuscu, I., 1993, Surface faulting on the North Anatolian Fault in these two Millennia: U.S. Geological Survey Open-File Report 94–568, p. 143–144, http://home.hiroshima-u.ac.jp/kojiok/anatolia.htm.

Palmer, A.R., 2000, Engaging "my neighbor" in the issue of sustainability, Part II: The context of humanity: Understanding deep time: GSA Today, v. 10, no. 2, p. 10.

Parsons, T., Toda, S., Stein, R., Barka, A., and Dieterich, J., 2000, Heightened odds of large earthquakes near Istanbul: An interaction-based probability calculation: Science, v. 288, p. 661–665, doi: 10.1126/science.288.5466.661.

Pimentel, D., Harvey, C., Resosundarmo, R., Sinclair, K., Kurz, D., McNair, M., Cris, S., Shpritz, L., Fitton, L., Saffouri, R., and Blair, R., 1995, Environmental costs of soil erosion and conservation benefits: Science, v. 267, p. 1117–1123, doi: 10.1126/science.267.5201.1117.

Pimm, S.L., Russell, G.T., Gittelman, J.I., and Brooks, T.M., 1995, The future of biodiversity: Science, v. 269, p. 347–350, doi: 10.1126/science.269.5222.347.

Posner, R.A., 2004, Catastrophic risk and response: New York: Oxford University Press, 322 p.

Rampino, M.R., and Self, S., 1992, Volcanic winter and accelerated glaciation following the Toba super-eruption: Nature, v. 359, p. 50–52, doi: 10.1038/359050a0.

Rampino, M.R., and Self, S., 1993, Climate-volcanism feedback and the Toba eruption of ~74,000 years ago: Quaternary Research, v. 40, p. 269–280, doi: 10.1006/qres.1993.1081.

Reitan, P., 2005, http://ejournal.nbii.org/archives/vol1iss1/communityessay.reitan.html.

Riley, S., Fraser, C., Donnelly, C., Ghani, A., Laith, J., Abu-Raddad, L., Hedley, A., Leung, G., Ho, L., Lam, T., Thach, T., Chau, P., Chan, K., Lo, S., Leunt, P., Tsang, T., Ho, W., Lee, K., Lau, E., Ferguson, N., and Anderson, R., 2003, Transmission dynamics of the etiological agent of SARS in Hong Kong: Impact of public health interventions: Science, v. 300, p. 1961–1966, doi: 10.1126/science.1086478.

Sagan, C., 1994, Pale Blue Dot: A Vision of the Human Future in Space: New York, Random House, 429 p.

Speth, J.G., 2004, Red sky at morning: America and the crisis of the global environment: New Haven, Yale, 299 p.

Stanley, D.G., and Warne, A.G., 1998, Nile delta in its destruction phase: Journal of Coastal Research, v. 14, p. 794–825.

Tainter, J., 1988, The collapse of complex societies: Cambridge, UK, Cambridge Univ. Press, 250 pp.

Tainter, J., 1995, The sustainability of complex societies: Futures, v. 27, p. 397–404, doi: 10.1016/0016-3287(95)00016-P.

Vallance, J.W., and Scott, K.M., 1997, The Osceola mudflow from Mount Rainer: Sedimentology and hazard implications of a high clay-rich debris flow: Geological Society of America Bulletin, v. 109, p. 143–163.

Wackernagel, M., and William R., 1996, Our ecological footprint—Reducing human impact on Earth: Gabriola Island and Stony Creek, New Society Publ., 160 p., see p 85, Table 3.4 (for 1991 data).

White, L., Jr., 1967, The historical roots of our ecologic crisis: Science, v. 155, p. 1203–1207, doi: 10.1126/science.155.3767.1203.

Wright, R., 2004, A short history of progress: Toronto, Anasi Press, 211 pp.

Yamaguchi, D.K., Atwater, B.F., Bunker, D.E., Benson, B.E., and Reid, M.S., 1997, Tree-ring dating the 1700 Cascadia earthquake: Nature, v. 389, p. 922–923, doi: 10.1038/40048.

Zen, E., 2000, Stakes, options, and some natural limits to a sustainable world, in Schneiderman, J.S., ed., The Earth around us: Maintaining a livable planet: New York, W.H. Freeman, p. 386–397, 440–441.

Zen, E., Barton, P.B., Reitan, P.H., Kieffer, S.W., and Palmer, A.R., 2002, Earth resources: The little engine that could brake sustainability: published on the Boulder Community Network site, http://bcn.boulder.co.us/basin/local/sustain_update.html.

MANUSCRIPT ACCEPTED BY THE SOCIETY 3 NOVEMBER 2008